THE POWER OF MAPS

MAPPINGS: Society / Theory / Space

A Guilford Series

THE POWER OF MAPS
Denis Wood (with John Fels)

APPROACHING HUMAN GEOGRAPHY
An Introduction to Contemporary Theoretical Debates
Paul Cloke, Chris Philo, and David Sadler

Forthcoming

POSTMODERN CONTENTIONS
Epochs, Politics, Space
John Paul Jones III, Wolfgang Natter, and Theodore Schatzki, Editors

The Power of Maps

by

DENIS WOOD

with John Fels

THE GUILFORD PRESS
New York London

In memoriam
Brian Harley

A Division of Guilford Publications, Inc.
72 Spring Street, New York, N. Y. 10012

Printed in the United States of America

This book is printed on acid-free paper.

Last digit is print number: 9 8 7 6 5 4 3 2

Library of Congress Cataloging-in-Publication Data

Wood, Denis.
The power of maps / by Denis Wood with John Fels.
p. cm. — (Mappings)
Includes bibliographical references and index.
ISBN 0-89862-492-4 — ISBN 0-89862-493-2 (pbk.)
1. Maps. I. Fels, John. II. Title. III. Series.
GA105.3.W66 1992
912—dc20 92-23443
CIP

Preface

Not all of this book is published here for the first time. Most of the substance of Chapters Five and Six and other material throughout the book appeared as articles, reviews and review essays in *Cartographica,* that essential organ without which thought in cartography today is hard to imagine. Thanks is only a tithe what is due its moving spirits, Bernard Gutsell and Ed Dahl, for their ability to see the future, encourage debate, and . . . *publish it.* Portions of Chapter Six also appeared in *Prologue: The Journal of the National Archives*, and I have also culled from work published in the *Bulletin* of the Geography and Map Division of the Special Libraries Association, *Environment and Behavior*, and the *Journal of Environmental Psychology* (high praise to David Canter too). Most of the ideas were worked out in endless conversations with Bob Beck and George McCleary, Arthur Krim and Tom Koch (who also copyedited the manuscript).

The original version of Chapter Five was coauthored with John Fels. I also plundered work he had completed on a collaboration in progress for material in Chapters One and Three. But indeed his thought is all over this book and I can scarcely imagine having written this had it not been for what I learned from him while working on "Designs on Signs." It was with John that I learned to understand Roland Barthes, whose insight and example informs, I would like to imagine, every page.

Peter Wissoker first suggested I might want to write this book. He also provided the stimulus that made it happen. Andrew Pekarik, Assistant Director of Programs, gave me the opportunity to draft it in the form of an exhibition of the same title at the Cooper-Hewitt Museum. My co-curator, Lucy Fellowes; the Curator of Exhibitions, Dorothy Globus; and the project researcher, Griselda Warr also made important contributions to the form this book took. That they knew my name I owe to Roger Hart. That they were willing to trust my judgment I owe to the example of the exhibition I curated in 1990 for the Brattleboro Museum of Art. The support there of Mara Williams, Linda Rubinstein and Sherry Bartlett made that show possible; and its example was essential to my being able to imagine the Cooper-Hewitt's *Power of Maps* and hence this book. That *they* knew my name I also owe to Roger Hart.

Robert Rundstrom organized the special sessions of the Association of American Geographers in San Diego this year where I was able to try out Chapter Two on an appreciative audience. George Thompson gave me access to material vital to the project, as did Ed Dahl. The aid of Cindy Levine and Eric Anderson, Dick Wilkinson and Bernie May, Mark Ridgley and Susan Goodmon, Tina Silber and Marc Eichen, and Gordie Hinzmann was also invaluable. Mr. Partridge got me everywhere on time. Martie Phelps typed Chapters Two, Five and Six. The book would never have been finished without her help. Thanks, Martie!

Ingrid made sure nothing extraneous got in my way while writing this; Randall was always on call to save me from my ineptitude at the computer (and he helped proofread the manuscript); Chandler never failed to ask, "How many pages now, Den?"; and Kelly did his damnedest to give me the peace of mind I needed to bring it in on time. Had I not felt compelled to dedicate this book to a man whose example I never adequately acknowledged during his lifetime, I would have offered it to these four whom I love.

Raleigh, N.C.
August 1992

Contents

Introduction

Power is the ability to do work. Which is what maps do: *they work.*

They work in at least two ways. In the first, they operate *effectively*. They work, that is . . . *they don't fail*. On the contrary, they succeed, they achieve effects, they get things done. *Hey! It works!* But of course to do this maps must work in the other way as well, that is, *toil*, that is, *labor*. Maps sweat, they strain, they apply themselves. The ends achieved with so much effort? The ceaseless reproduction of the culture that brings them into being.

What do maps *do* when they work? They make present—they *re*present—the accumulated thought and labor of the past . . . about the milieu we simultaneously live in and collaborate on bringing being. In so doing they enable the past to become part of our living . . . *now* . . . *here*. (This is how maps facilitate the reproduction of the culture that brings them into being.) The map's *effectiveness* is a consequence of the *selectivity* with which it brings this past to bear on the present. This selectivity, this focus, this particular attention, this . . . *interest* . . . is what frees the map to be a *re*presentation of the past (instead of, say, a time machine, which because it makes the past wholly present eliminates . . . *the present*). It is this interest which makes the map a representation. This is to say that maps work . . . by serving interests.

Because these interests select what from the vast storehouse of knowledge about the earth the map will represent, these interests are embodied in the map as presences and absences. Every map shows *this* . . . but not *that*, and every map shows what it shows this *way* . . . but not *the other*. Not only is this inescapable but it is precisely because of this interested selectivity—this choice of word or sign or aspect of the world *to make a point*—that the map is enabled to work.

This interest is not disinterested. Neither is it simple or singular. This interest is at once diffused *throughout* the social system (in those forces enabling the reproduction of the society as a whole) and *concentrated* in this or that moiety, guild, or class, in this or that gender, occupation, or profession, in this or that neighborhood, town, or country, in this or that . . . *interest*. Rhetorical expedience dictates a certain . . . *reticence* with respect to both the site of the interest and its particular goals and aims. Therefore the conservative forces, whose end is social reproduction in general (and, let us face it, the position of those dominant in it), and the transformative forces, attendant to the interests of this or that particular class or industry or part of the country, *conspire* to mask their interest, conspire to . . . *naturalize* this product of so much cultural energy.

This *naturalization of the map* takes place at the level of the sign system in which the map is inscribed. A double coding ensures this (it is at least a double coding). No sooner is a sign created than it is put to the service of a myth (this is that the world displayed in the map is . . . *natural*). It is thus not merely that the native Americans were left off maps made by Europeans in the 16th and 17th centuries, but that the resulting surface—of trees, rivers, hills—took on the appearance of a window through which the world was seen . . . *as it really was*.

The map's ability to do this depends on the signs it deploys, all of which have . . . independent histories. These suggest not only how interests are embodied in individual signs, but how the signs themselves are brought into being and reproduced. But with maps revealed as the historically contingent sign systems they in fact are, they cease to appear as miraculous windows through which to snap shots of reality. No longer confused with the world, maps are suddenly capacitated as powerful ways of making statements about the world.

Maps are finally enabled to work . . . *for you*.

* * *

This thesis is embodied in this book in seven chapters. The first attempts to say how it is that maps work, how they make present—so that we can use it today—the accumulated thought and labor of the past.

The second strives to embed the map in the history out of which it emerges, as an instrument that serves interests which not only bring maps into being, but in so doing divide the world . . . between those who use maps and those who don't.

The third tries to show the way that these interests are embodied . . . in *every* map (even the most *objective seeming*) though a close reading of Tom Van Sant's brilliant map of the world.

The fourth pursues the way the map *masks* the interest it embodies—the way it *naturalizes* the map—through a close reading of the topographic survey of the area around Ringwood in northeastern New Jersey.

Through a close reading of the *Official State Highway Map of North Carolina* the fifth chapter tries to demonstrate the way this mask is embedded in the very signs either into which the map may be decomposed or out of which it can be constructed.

In the sixth chapter the history of a single one of these signs—that for landform relief (that is . . . *hills*)—illuminates the continuity of the map with the rest of the culture, the way its interest . . . is the map's.

The seventh chapter insists that the realization of the interest every maps serves frees it to serve . . . mine, yours; that is, to work . . . *for us*.

CHAPTER ONE

Maps Work by Serving Interests

A cornucopia of images, bewildering in their variety: this is the world of maps. Sticks and stones, parchment and gold leaf, paper and ink . . . no substance has escaped being used to frame an image of the world we live in. Like the birds and bees we have danced them in the gestures of our living; since the birth of language we have sketched them in the sounds of our speech. We have drawn them in the air and traced them in the snow, painted them on rocks and inscribed them on the bones of mammoths. We have baked them in clay and chased them in silver, printed them on paper . . . and tee-shirts. Most of them are gone now, billions lost in the making or evaporated with the words that brought them into being. The incoming tide has smoothed the sand they were drawn in, the wind has erased them from the snow. Pigments have faded, the paper has rotted or been consumed in the flames. Many simply cannot be found. They are crammed into the backs of kitchen drawers or glove compartments or mucked up beneath the seats with the Kentucky Fried Chicken boxes and the paper cups. Where have all the road maps gone: and the worlds they described and the kids we knew, Route 66, and the canyon beneath Lake Powell, and the old Colorado pouring real water into the Gulf of Mexico? And when we talk of the "old map of Europe"—which too has disappeared—we are speaking of certainties we grew up with, not a piece of paper. And yet, and yet . . . it is hard, in the end, to separate those certainties from that very piece of paper which not only described that world, but endowed it with a *reality* we have all accepted.

A Reality Beyond Our Reach

And this, essentially is what maps give us, *reality*, a reality that exceeds our vision, our reach, the span of our days, a reality we achieve no other way. We are always mapping the invisible or the unattainable or the erasable, the future or the past, the whatever-is-not-here-present-to-our-senses-now and, through the gift that the map gives us, transmuting it into everything it is not . . . *into the real.* This month's *Life* leaps at me from the checkout counter: "Behold The Earth," it says. "Startling new pictures show our planet as we've never seen it before." Inside, below the heading "This Precious Planet," the copy promises "Striking new views from near space show us more than we could ever have guessed about our fragile home."[1]

Outside in the parking lot I am not struck by the preciousness of the planet, much less by its fragility. Instead, I am overwhelmed by the solidity and apparent indestructibility of everything I see around me. Only the pictures—let us think about them as maps for the moment—convince me of the reality the captions evoke. "Behold the Earth": it is as if we had never done so before, and indeed . . . *apparently we haven't.* "New pictures"; "never seen it before"; "new views"; "show us more": each phrase insists on the fact that indeed I *never have seen* the planet in quite this way.

Let's face it: I haven't. Neither have you. Few have. At most even the best traveled have seen but a few square miles of its surface: the space around this convention center, that neighborhood, the thin traverse of the tour bus, the road from the airport home. It is not ample, this territory we individually occupy. It scarcely deserves the name "world" much less "planet." I think of what Arthur Miller wrote about his father:

> In his last years my father would sit on the porch of his Long Island nursing home looking out on the sea, and between long silences he would speak. "You know, sometimes I see a little dot way out there, and then it gets bigger and bigger and finally turns into a ship." I explained that the earth was a sphere and so forth. In his 80 years he had never had time to sit and watch the sea. He had employed hundreds of people and made tens of thousands of coats and shipped them to towns and cities all over the States, and now at the end he looked out over the sea and said with happy surprise, "Oh. So it's round!"[2]

Why should it be otherwise? The sphericity of the globe is not something that comes to us as seeing-hearing-sniffing-tasting-feeling animals, is not something that comes to us . . . *naturally*. It is a residue of cultural activities, of watching ships come to us up out of the sea *for eons*, of thinking about what that might mean, of observing shadows at different locations, of sailing great distances, of contemplating all this and more at one time. It is hard won knowledge. It is *map* knowledge. As such it is something that little kids have to learn, not something they can figure out for themselves. "Educators are living in a dream world if they assume young children understand that the earth is round," write Alan Lightman and Philip Sadler. Even many fourth graders who say the earth is round often "picture a flat part where people live in the interior of the ball. Others draw the earth as a giant pancake or as a curved sky covering a flat ground."[3]

Even *these* images—*these* maps—exceed the raw experience of the kids, are informed and supported by the cultural activities that inform and support mapping: knowledge, graphic conventions, ideas about representation, conventional ways of conceptualizing earth and sky and our place between them.

So how *do* we know the earth is round? We know the earth is round because (almost) everybody says it's round, because in geography class our teachers tell us it is round, because it is round on map after map after map . . . or, if not precisely round, then *supposed* to be round, topologically round, so that when you run your finger off one side of the map, you have the license to put it back down on the other. This is not some form of solipsism, but an effort to understand why in so many media we have made so many maps for so many years. Ultimately, the map presents us with the reality we *know* as differentiated from the reality we see and hear and feel. The map doesn't let us *see* anything, but it does let

Hedy Ellis Leiter, age 7, draws the world.

us know what *others have seen* or found out or discovered, others often living but more often dead, the things they learned piled up in layer on top of layer so that to study even the simplest-looking image is to peer back through ages of cultural acquisition.

Here, another image from this month's *Life*, one of Pacific winds. It is probably less than three square inches on the page, a voluptuous circle of swirling color summarizing . . . *millions of pieces of data*. And if the caption makes reference only to the computer display of satellite transmissions, *we* can see in its implied sphericity the Greeks and the Chinese who pondered long the meaning of the ships coming *up* from the sea; we can set sail once more with Columbus and Magellan, stand again upon a peak in Darien and stare out with Cortez at the Pacific; we can walk the decks of the ships and ride the buoys that used to make these measurements; we can . . . take advantage of *all the work* that has gone before, all the ingenuity and effort, all the voyages taken and flights made, all the hypotheses advanced and demolished and finally proven, all caught, all taken advantage of, *all justified*, by this silver-dollar sized hot-pink and blue map of Pacific winds.

Maps Make the Past and Future Present

The world we take for granted—the real world—is made like this, out of the accumulated thought and labor of the past. It is presented to us on the platter of the map, *presented*, that is, *made present*, so that whatever invisible, unattainable, erasable past or future can become part of our living . . . *now* . . . *here*. An example: I am one of a group of Raleigh citizens who have banded together to oppose a road the City of Raleigh wants to build across the grounds of a hospital listed on the National Register of Historic Places. In the process—of our living here and now—we compare a map of the proposed route for the road—that is, a map of a potential future—with a map of the historic site—that is, with a map displaying a determination made in the past about the extent of the historic site. Past and future—neither accessible to my senses *on the ground* (the road does not yet exist, there is nothing to see, the boundaries of the historic district are not yet inscribed in the dirt, there is not even a marker)—come together in my present through the grace of the map.

And every map is like this, every map facilitates some living by virtue of its ability to grapple with what is *known* instead of what is merely seen, what is *understood* rather than what is no more than sensed. I want to say that recently the distance between this visible, palpable world of our senses and the world we make of it has stretched. On the cover of Stephen Hall's *Mapping the Next Millennium*[4] is what appears to be a map of the ocean floor. Actually it displays anomalies in the gravity field of the ocean floor based on radar altimeter measurements of the sea surface that mimic the topography of the ocean floor. Now, that's a stretch from the sandy bottom beneath our feet, or even from the (now) old-fashioned sonar readings we used to have to rely on. But here's another map, the first in *The New State of the World Atlas*.[5] It looks like a political map of the world, and that's what it is. But just because it's easy for us to say this doesn't mean that conceptually it's not a stretch. Once you start to think about it, you realize that conceptually it's a lot more of a stretch than the map of the ocean floor. In fact, once you start to think about it, you realize that it's very difficult to say what this is actually a map of or to describe how it came to be. The concept of a gravity anomaly may or may not pose a conceptual difficulty, but the idea of a satellite bouncing radar off the ocean surface to map subtle variations in its height is straightforward; and the idea that these variations might reflect subtle variations in the gravitational field of the ocean floor that might in turn relate to variations in its topography is not too convoluted either. One can imagine the sensing system, can cope with the idea of its data being turned into this image.

But with the political map, this straightforward quality vanishes. National boundaries are not sensible. If variations in land use (as between Haiti and the Dominican Republic), or the gauge of railroad track (as between Russia and China), or the orientation of mailboxes (as between Vermont and Quebec), indicate the presence of an otherwise insensible border, no less often there is no difference to mark such a boundary through the rain forest (between Bolivia and Brazil), or across the desert (between Oman and Saudi Arabia), or in Los Angeles (between Watts and Compton). Or, the opposite situation, there *is* a chain link fence dripping with concertina wire and guard posts establishing the rhythm of a certain paranoia, and this border, which is more than sensible, *is not the border*, the border is contested, the neighbors disagree, there are binding United Nations' resolutions that are ignored, atlases show the border . . . *somewhere else*.[6] Here the stretch between the sensible and the mapped is close to the breaking point: what *is* being mapped?

Every map constitutes such a stretch, those of the big world no more than that of the lot our house sits on, whose description reads as follows on the deed to our property:

> Beginning at a stake marking the northeastern corner of the intersection of West Cabarrus Street and Cutler Street and running thence along the eastern line of Cutler Street North 3° 17' West 50 feet to a stake, the southwestern corner of Lot 125 as shown on map reference to which is hereinafter made; runs thence [*and so forth and so on*] to the place and point of Beginning, and being Lot Number 126 of Boylan Heights according to map recorded in Book of Maps 1885, page 114, Wake County Registry.

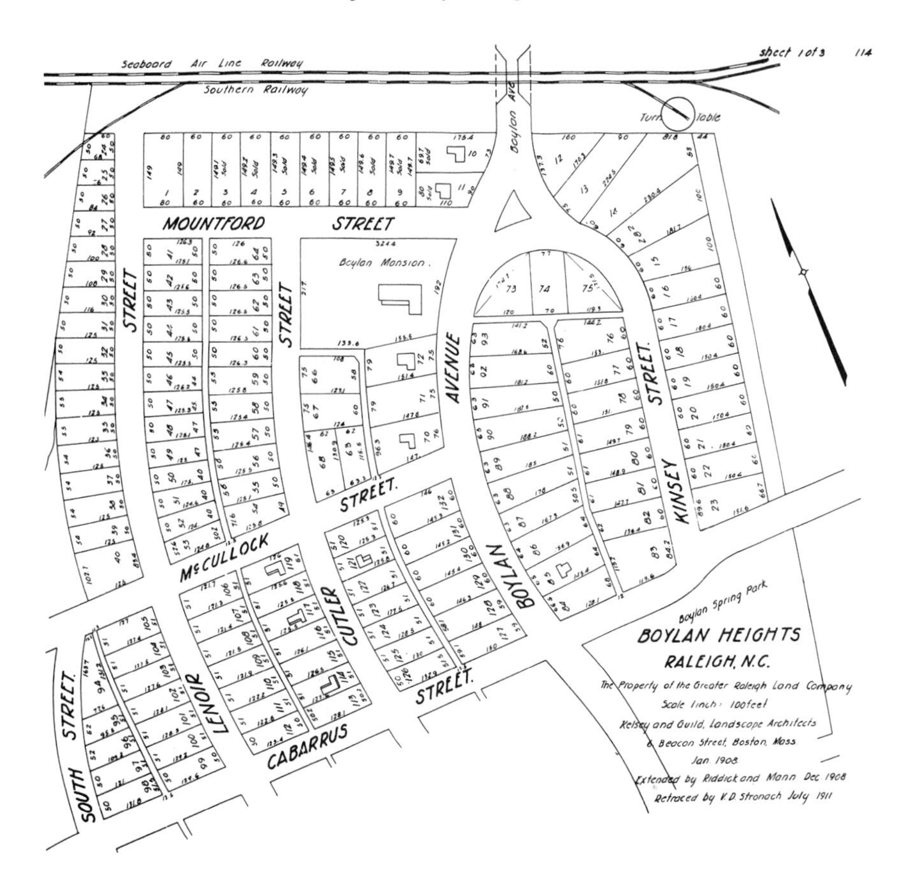

The map, from the *Book of Maps 1885* (p. 114), Wake County Registry, on which lot number 126 is recorded.

But there is no stake, there are no stakes, there is nothing to see; or where there is, all acknowledge that the fence does not follow the property line but veers across it; the only reality is the map, the map recorded on page 114 of the Book of Maps 1885. Here is the stretch—there is nothing in the trees or grass, on the sidewalk or street to mark the ownership the map grants us (the *land* is there: it is the *property* the map creates)—and here again is this *activity of another world*—the past in which control of this land was seized by the English Crown and granted to those who sold or gave it to those who sold or gave it to those who sold or gave it to those who sold it to us—made present in the map so that it could be made part of our living . . . *here* . . . *now*.

How does the map do this?

It does it by connecting us through it to other aspects of a vast system similarly brought forward from the past and embodied, not in maps, but in codes, laws, ledgers, contracts, treaties, indices, covenants, deals, agreements, in pledges, in promises, in words given and oaths taken. Through this map, for instance, the ownership of the property it grants us—whose limits it describes, whose limits it makes real for us—is tied to a hierarchy of tax codes. The owners of Lot 126—described in Book of Maps 1885, page 114—have obligated themselves, through their purchase, to pay taxes to the county (itself a creature of another set of maps). Through these in turn they have linked themselves to the local school district (endowed with reality by yet another map) where their children attend school (in an attendance zone defined by still another map). Through their purchase, they have similarly obligated themselves to observe a set of restrictions on the use of their property that are embodied in zoning maps (they cannot rent out, for instance) as well as in an historic district overlay (they must receive approval from an appearance commission before they can paint their house any color other than white). Others, connected to the owners of Lot 126 through their own enmeshment in this hierarchy of nested maps, have identical *and* reciprocal obligations. They have agreed not to dump trash on Lot 126, or set their pup tents on it, or use it for a playing field, or as a shortcut; they have agreed to help pay for the garbage collection from the alley behind the Lot, and to help pay for the water and sewerage, fire and police protection, and other services . . . *that come with the territory.*

Maps Link the Territory with What Comes with It

It is this ability *to link the territory with what comes with it* that has made maps so valuable to so many for so long. Maps link the territory with taxes, with military service or a certain rate of precipitation, with the likelihood that an earthquake will strike or a flood will rise, with this or that type of soil or engineering geology, with crime rates or the dates of first frost, with parcel post rates or area codes, with road networks or the stars visible on a given date. Maps link land with all these and with whatever other insensible characteristics of the site past generations have been gathering information about for whatever length of time. The University Museum at the University of Pennsylvania has a property map subserving some of these functions that is three thousand years old. It was incised with cuneiform characters on a clay tablet in Mesopotamia,[7] but no subsequent society of any size has long failed to make property maps in a variety of media. Ancient Egyptians drew them and Roman *agrimensores* surveyed them; the Japanese had them made as long ago as 742 AD, and there is an Aztec map of property ownership in the Library of Congress dated to 1540.[8] With the passage of the Land Ordinance of 1785 in the United States, and the cadastral mapping of France set in motion 1807 by Napoleon, increasingly enormous swaths of the planet were entered into this huge atlas of proprietorship, until now it is hard to imagine there is a square inch left whose ownership has not been staked out, squabbled over, bought, sold or killed for, each *transaction* . . . recorded someplace, on a map in a land office.

Such maps account for but a single layer in the great bundling of boundaries with which we have tied up the planet: maps of treaty organizations and national borders; maps of provinces, territories and states; maps of boroughs, counties, parishes and townships; maps of towns and cities, neighborhoods and subdivisions; maps of water and soil conservation districts; maps of garbage collection routes and gas service districts; fire insurance and land-use maps; precinct maps; tithing maps; congressional district maps; maps of the jurisdiction of courts . . . There is no reason to end this list here—*or anywhere*—for there are as many different kinds of—what to call these? boundary maps? power projection maps?—as there are ways of holding sway upon the earth.

And such boundary maps constitute but a single entry in the vast ledger maps keep. To open any thematic atlas is to—here, these are the plate titles, that is, the names of the things in the world the maps point to, from the "world thematic maps" section of *Goode's World Atlas:* Political, Physical, Landforms, Climatic Regions, Surface Temperature Regions, Pressure, Winds, Seasonal Precipitation, Annual Precipitation, Ocean Currents, Natural Vegetation, Soils, Population Density, Birth Rate, Death Rate, Natural Increase, Urbanization, Gross National Product, Literacy, Languages, Religions, Calorie Supply, Protein Consumption, Physicians, Life Expectancy, Predominant Economies, Major Agricultural Regions, Wheat, Tea, Rice, Maize, Coffee, Oats, Barley—well, it goes on for pages.[9] Or here, a *totally* different selection from *The New State of War and Peace* atlas: The Dove of Peace (a map of cease-fires and reductions in armed forces, 1988–90), The Dogs of War (a map of states in which wars took place, 1989–90), Unofficial Terror, Nuclear Fix, Killing Power, The Killing Fields, Bugs and Poisons, The Armourers, The Arms Sellers, The Butcher's Bill (the number of deaths attributable to war), The Displaced (the number of refugees), Sharing the Spoils, The Martyred Earth . . . and this too goes on for pages.[10] Zoom in? In the 57 maps of *The Nuclear War Atlas* we can subject the Nuclear Fix map of the *The New State of War and Peace* atlas to a kind of microscopic inspection. For example, here is a map showing the destruction of Hiroshima during World War II, and here another showing the sweep of debris around the world from the fifth Chinese nuclear detonation, and here a third showing the portions of the United States that would receive more than 100 rems of radiation in a nuclear war. There are 54 more where these came from.[11]

Zooming out allows us to take in what cartographers refer to as general reference maps, images establishing a relatively indiscriminate reality, at least by the standards of the *Nuclear War Atlas*, the *Atlas of Landforms*[12] or *The World Atlas of Wine*.[13] In *Goode's* these comprise most of the maps in its "regional section" where they go under the name of "physical–political reference maps," that is, maps that pay attention to selected aspects of the physical environment—topography and major water features—and to a few of what are called "cultural features"—political boundaries, towns and cities (of certain sizes), roads, railroads, airports, dams, pipelines, pyramids, ruins and . . . *caravan routes*.[14] Granted . . . it's hardly *general* reference, but it's about as close as maps come to portraying a world that we might *see*, especially at really large scales where, when the relief is shaded, the maps begin to suggest *pictures* of the world as it might be seen from an airplane . . . sort of. The map is always a stretch. It is never "the real thing" we walk on or smell or see with our eyes:

> Big Tiger had never had a map in his hand before, but he pretended to know all about maps and remarked airily: "I can't read the names on this one because they're in English." Christian realized he would have to show his friend how to read a map. "The top is north," he said. "The little circles are towns and villages. Blue means rivers and lakes, the thin lines are roads and the thick one railways." "There's nothing at all here," said Big Tiger, pointing to one of the many white patches. "That means it's just desert," Christian explained "You have to go into the desert to know what it looks like."[15]

Exactly. This is the very point of the map, to present us not with the world we can *see*, but to point *toward* a world we might *know:*

> "That's a fine map," said Big Tiger. "It's useful to be able to look up beforehand the places we reach later." "Are there really bandits about here?" asked Christian. "Perhaps it's written on the map," Big Tiger ventured. "Look and see."[16]

And if caravan routes . . . *why not bandits?*

Maps Enable Our Living

Here is the difference between a property map and a general reference map: one nails us to the territory, the other merely points it out. We might use a property map (and the maps to which it points) to answer *beforehand* questions about school districts and crime rates, but the way such a map usually *works* is to make these connections effective in the *ongoingness* of our daily living. The general reference map . . . *is a less involved observer*. The difference is that between parents saying "Wash these!" as they point to the dirty dishes—the property map—and saying

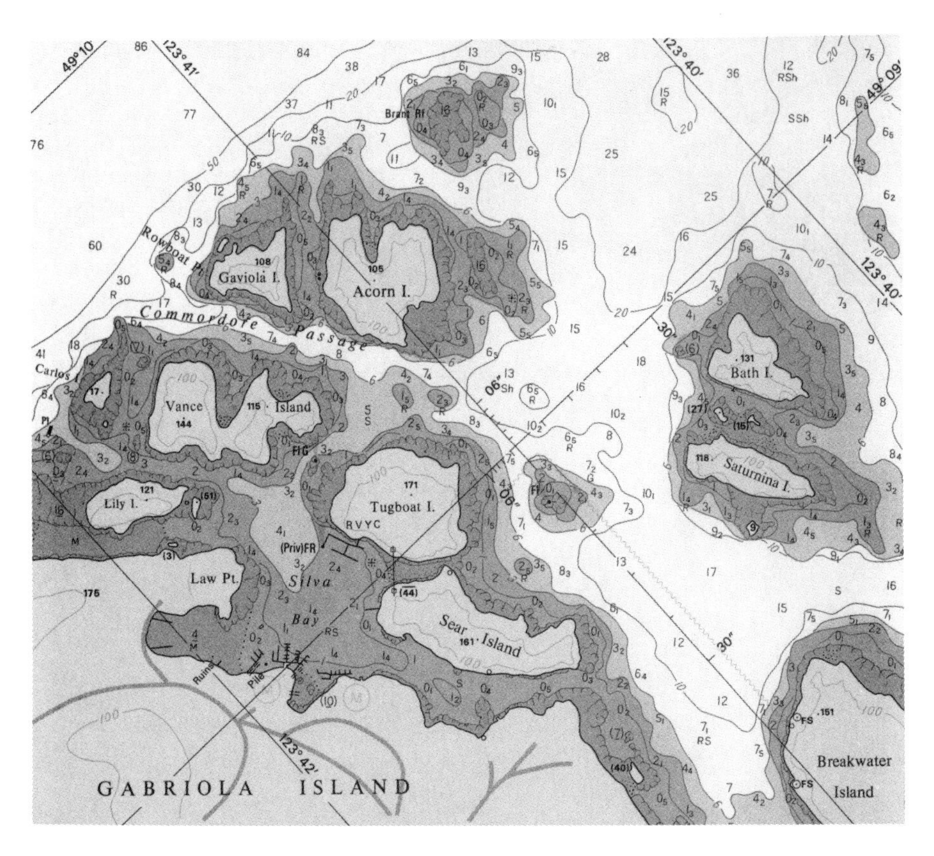

A detail from "Porlier Pass to Departure Bay," the fourth of four sheets which together make up *Chart 3310—Gulf Islands: Victoria Harbour to Nanaimo Harbour.*

"Your grandmother gave us those" as they wave in the general direction of the china cupboard. The indexicality is none the less, but the way that other world is made part of our living is less *well defined*, is less *enforceable*. This suggests that we can distinguish among maps not merely on the basis of what they show, but on the basis of the different livings into which that knowledge is incorporated. Spread before me is "Porlier Pass to Departure Bay," the fourth of four sheets that taken together comprise "Chart 3310—Gulf Islands: Victoria Harbour to Nanaimo Harbour."[17] It's a skinny chart about four feet long by maybe a foot and a half high, folded, meant to be used—that is, *consulted*—in sections, though when unfolded it's pretty enough, and more than one of these is hanging behind glass somewhere, decorating a wall *and* making a connection to the territory. The sheet is sprinkled with black numbers and furrowed with blue lines indicating the depth of the water in fathoms (under 11 fathoms in fathoms and feet). A cabala of other marks differentiates 46 aids to navigation (permanent lights, whistle buoys, fog signals), eight different qualities of seabed (gravel, mud, shells), and 44 other objects of interest (drying rocks with heights, kelp, wrecks, abandoned submarine cables).[18] Now, the living into which all of the labor that resulted in the production of this chart is incorporated is different from the living taking advantage of another map, say, this one, the *Geologic Map of Region J, North Carolina,*which shows in black and red the location of igneous, sedimentary, metasedimentary and metavolcanic rocks, various strikes and dips, and mineral resources, including, among other things, sites of potentially economic mineral deposits of crushed stone, iron and ruby corundum. Indicative of the living into which this map might be incorporated is the title of the booklet it accompanies, *Region J Geology: A Guide for North Carolina Mineral Resource Development and Land Use Planning.*[19] This allows us to imagine planners—as opposed to pilots—consulting this map in their struggle to make land-use decisions, as, for example, where to locate a low-level radioactive waste dump.

Conventionally we have been asked to think about these as different *uses* of maps—navigation, planning—but both exploit the maps' inherent indexicality to link the territories in question with what comes with them, here perhaps shoal water, there perhaps an active fault. The *uses* are less different than the *livings* that incorporate into their present the endless labor all maps embody. This is what it means to use a map. It may look like wayfinding or a legal action over property or an analysis of the causes of cancer, but always it is this incorporation into the here and now of actions carried out in the past. This is no less true when those actions are carried out . . . *entirely in our heads:* the maps we make in our minds embody experience exactly as paper maps do, accumulated as we have made our way through the world in the activity of our living. The deep experience we draw upon, for example, whenever, we select from the myriad possibilities *this route* for our trip to the movies is no less a product of work than was a medieval *portolan*, incorporating as it did in its making the accumulated knowledge of generations of mariners (and others) in the carefully crafted web of rhumb lines, the fine details of the coasts.[20] Onto the simple schemata with which we came into the world, our early suckling and crawling and grasping and peek-abooing all mapped a web of simple topological relations. This provided a substrate for the etching—as we moved out into the school yard and the neighborhood, as we explored the woods behind grandma's house or the meadows down beyond the creek—of spatial relations invariant under changes in point of view. Once we coordinated these, we could begin the construction of systems of reference invariant under changes in location, we could begin making . . . *maps*,[21] which we do, wherever we go, whenever we go, out of our movement on foot and in car, in boat and in plane, out of pictures we see and movies we watch, out of the things we read in the newspapers and hear on the radio, out of the books we read, the maps we consult, out of the atlases we flip through . . . out of the globes we spin. It is all labor, it is all work, the construction of these mental realms; and when we draw on them—for even the most mundane activity—we are bringing forward into the present this wealth we have laid up through the sweat of our brows.

To what end? To the same end, to—through the map—link all this elaborately constructed knowledge up with our living. Say we want to go to a movie. What do we do? We look it up in the newspaper, and read "Six Forks Station (daily at 2, 4:30, 7 and 9:30). Tower Merchants (nightly at 7, 9:30)." To choose which theater to go to, *much less how to get there*, we have to organize all the relevant bits of information into some kind of structure. For the moment let's call this structure a mental map, and let's think about it as a board sort of like a Paris metro map but covered with a trillion tiny light bulbs. When I think of Six Forks Station, a string of these bulbs lights up. This isn't the image I get, of course, any more than the activity of the computer I'm typing this on displays for me the machine processes in which it is engaged. The string of bulbs lights up, and I have a sense of Raleigh and a route to the theater, where it is (does my body sort of turn toward it?), what kind of roads will take me there (there may be many alternatives), the level of traffic at the time we want to go, other things. I don't know if this sense is displayed in my head as a map *image*. I know I can externalize it this way, but my feeling is that the mental maps I consult are less . . . *straightforward*. Sometimes the string won't be complete, I'll sense a gap in my knowledge, a little uncertainty, I'll say, "Do you know how to get there?" "Don't you?" "Umm . . . *sort* of." Maybe there's a red bulb at the end of one of the strings that makes me realize that I'll have to ask around when I get there, but that this won't be a problem. Or maybe there's a blue bulb that lets me know I could easily get lost. Of course this is only a map. There are no guarantees. I could get lost no matter the color bulb. The same thing happens for Tower Merchants. Among the alternatives for the two theaters, I select a couple to compare (do some routes glow more brightly than others?). Then the board goes black, and only these two routes light up again. One comes on in pink (heavy traffic), another in blue (road construction). In the end—this all takes milliseconds—I say, "What about Tower Merchants at 7?" and off we go.[22]

Of course a mental map is not a board with a bunch of lights on it, but the neurological activity underwriting this kind of decision is clearly related to the way we use paper maps to make decisions. Certainly the similarities increase once we begin to externalize these maps, to share them with each other. "What? Why would you go that way?" "Because it's shorter." "No, no, it's shorter if you take St. Mary's to Lassiter Mill—"

"Oh, and then go out Six Forks to Sandy Forks?" "Yeah." Here the maps, in separate heads, are being consulted almost as if they were paper maps open on the table, linking knowledge individually constructed in the past to a shared living unfolding in the present.

One Map Use—Many Ways of Living

For these many maps then, only one use (aside from swatting flies or wrapping presents in them), and that is this connecting up what-we-have-done (money we have exchanged, surveys we have carried out, walks we have taken) *through the map*—property or mental, thematic or general reference—with what we want to do or have to do, with what we find is pressing. But if only one use . . . *many livings*. Perhaps this is the problem with the many taxonomies of maps that have been attempted, they end up taxonomies not of maps, but of the ways we view the world and the many ways we make our way within it. Take this simple-looking scheme from *The Map Catalogue*.[23] Since its subtitle reads *Every Kind of Map and Chart on Earth and Even Some Above It*, we should be able to anticipate a certain . . . comprehensiveness. But what we find are three types of maps, of *land*, of *sky* and of *water*. This apparent simplicity—itself an illusion—disintegrates immediately. Under "Land Maps" are listed: Aerial Photographs, Agricultural Maps, Antique Maps, Bicycle Route Maps, Boundary Maps, Business Maps, Census Maps, CIA Maps, City Maps, Congressional District Maps, County Maps, Emergency Information Maps, Energy Maps, Foreign Country Maps, Geologic Maps, Highway Maps, Historical Site Maps, History Maps, Indian Land Maps, Land Ownership Maps . . . But already I'm exhausted with this inventory, there is no rhyme or reason to it, it is a melange, a potpourri . . . and it doesn't stop. Here, another one, this from a special issue of *The American Cartographer* (Journal of the American Congress on Surveying and Mapping) containing the *U.S. National Report to ICA, 1987*.[24] With all this we should be able to expect a certain . . . authoritativeness. Again, we have three fundamental divisions, but this time into *government mapping, business mapping* and *university cartography*. Again, the apparent simplicity is delusional (all the universities are state universities, under the latter we find "Limited Edition Maps for Corporate Cartography"), the divisions are not real, or they have to do with making money not maps, and, again, the whole dissolves into a chaos ordered only by the type on the page: "Cartographic Programs and Products of the U.S. Geological Survey," "NOAA Map and Chart Products," "Defense Mapping Agency Redesign Studies," "Maps for Parklands," "An Experimental 1:100,000 Ground/Air Product." A third example, this from the fifth edition of *the* textbook in the field, Robinson, Sale,

Morrison and Muehrcke's *Elements of Cartography*.[25] Again, we have three divisions: "In order to provide a basis for the appreciation of the similarities and differences among maps and cartographers, we will look at maps from three points of view: (1) their scale, (2) their function, and (3) their subject matter." There is the imputation that these are independent, but again the classification collapses on inspection, this time into . . . *the vague*. Scale turns out to distinguish between *the large* and *the small*. Function discriminates among *the general*, *the thematic* and *the chart*. Under subject matter—after a nod toward cadastral mapping and plans—we find that "there is no limit to the number of classes of maps that can be created by grouping them according to their dominant subject matter." This leads to the conclusion that "cartography is independent of subject matter," thus rendering moot the point of making it the basis of a classification in the first place.

Hall gives us *four* divisions (violent novelty)—"Planetary Landscapes, Ours and Others," "The Animate Landscape" (maps of the body, brain, gamete, genes and DNA), "Probabilistic Landscapes, Atomic and Mathematical" (atomic surfaces, particle interactions, the fractal mapping of *pi*), and "Astronomical and Cosmological Landscapes".[26] So does *Goode's*, though all four of *Goode's* —world thematic maps, major cities maps, regional sections and ocean floor maps[27]—would get lost in a single division of Hall's. Southworth and Southworth, both designers, give us eight in a veritable explosion of map types: Land Form; Built Form; Networks and Routes; Quantity, Density and Distribution; Relation and Comparison; Time, Change and Movement; Behavior and Personal Imagery; and Simulation and Interaction. Bizarrely enough, they refer to these as "mapping techniques," including what others call map types within them (thus: embossed map, relief map, route map, diagrammatic strip map, pictographic map, cartoon map, military map, geologic map, pictorial map, insurance map) but making no effort to systematize these.[28]

Maps Construct—Not Reproduce—the World

These disparate efforts have in common precisely what the maps they so desperately attempt to sort have in common. Both are driven and shaped by the uses that connect the maps *through* them—*through* the taxonomies—to the livings that demanded and produced them. The crude impulse to produce a book produced the crude taxonomy of the *The Map Catalogue* with its land, sky and water world arbitrarily decomposed according to the order of the letters in the alphabet. The cartography journal, written by and for people who *make* maps, followed the cleavages of production. Hall, a journalist on the prowl for "newly charted realms," found them in the sub- and supraterrestrial, and from these he generated a taxonomy of spectacle. The Southworths produced—as designers might be expected to—a more formal partitioning. But for all this, little taxonomy . . . *of maps*. One would not be difficult to imagine. Here: at the level of the kingdom, *material maps* and *mental maps*. Within the kingdom *material*, phyla distinguish among substances: *paper maps*, *cloth maps*, *clay maps*, *metal maps*. Within these, subphyla and classes, by size and weight; orders and families, by age and place of production; genus and species, by projection and . . . Well, at least it's a taxonomy of *maps*, instead of the earth or of mapmakers or the elements of Aristotle. But it is immediately evident how . . . *uninteresting* these classifications are, how . . . *irrelevant*. Not that the size and weight of the map don't matter—you can't read the *Times Atlas* in bed; a map you're going to steer by needs to fold up into small sections—but that these characteristics are subsumed in the more general, more powerful, more . . . *meaningful* question of how the map will link its readers to the world it embodies. Thus: bicycle map, blueprint, book illustration, topo sheet, historical atlas, wall map, logo . . . .

Again: caught in the net of the living. Better simply . . . *to admit it that knowledge of the map is knowledge of the world from which it emerges*—as a casting from its mold, as a shoe from its last—isomorphic counter-image to *everything* in society that conspires to produce it. This, of course, would be to site the source of the map in a realm more diffuse than cartography; it would be to insist on a *sociology* of the map. It would force us to admit that the knowledge it embodies was socially *constructed*, not tripped over and no more than . . . *reproduced*. But then no aspect of the map is more carefully constructed than the alibi intended to absolve it of this guilt. In his effort to understand why historians make so little use of maps, Brian Harley argues that it follows from the way they see them:

> The usual perception of the nature of maps is that they are a mirror, a graphic representation, of some aspect of the real world. The definitions set out in various dictionaries and glossaries of cartography confirm this view. Within the constraints of survey techniques, the skill of the cartographer, and the code of conventional signs, the role of a map is to present a factual statement about geographic reality. Although cartographers write about the art as well as the science of mapmaking, science has overshadowed the competition between the two. The corollary is that when historians assess maps, their interpretation is molded by this idea of what maps are supposed to be. In our own Western culture, at least since the Enlightenment, cartography has been defined as a factual science. The premise is that a map should offer a transparent window on the world.[29]

What is achieved in this way? Precisely the pretense that what the map shows us is . . . *reality*. Were it not reality, why then it would just be

. . . *opinion*, somebody's *idea* of where your property began and ended, a good *guess* at where the border was, a *notion* of the location of the hundred-year flood line, but not the flood line itself. What is elided in this way is precisely the social construction of the property line, the social construction of the border, the social construction of the hundred-year flood line, which—like everything else we map—is not a line you can *see*, not a high water mark drawn in mud on a wall or in debris along a bank, but no more than a more-or-less careful extrapolation from a statistical storm to a whorl of contour lines. As long as the map is accepted as a window on the world, these lines must be accepted as representing things in it with the ontological status of streams and hills.[30] But no sooner are maps acknowledged as social constructions than their contingent, their conditional, their . . . *arbitrary* character is unveiled. Suddenly the things represented by these lines are opened to discussion and debate, the *interest* in them of owner, state, insurance company is made apparent. Once it is acknowledged that the map *creates* these boundaries, it can no longer be accepted as *representing* these "realities," which alone the map is capable of embodying (profound conflict of interest).[31] The historian's problem is everybody's problem: our willing-

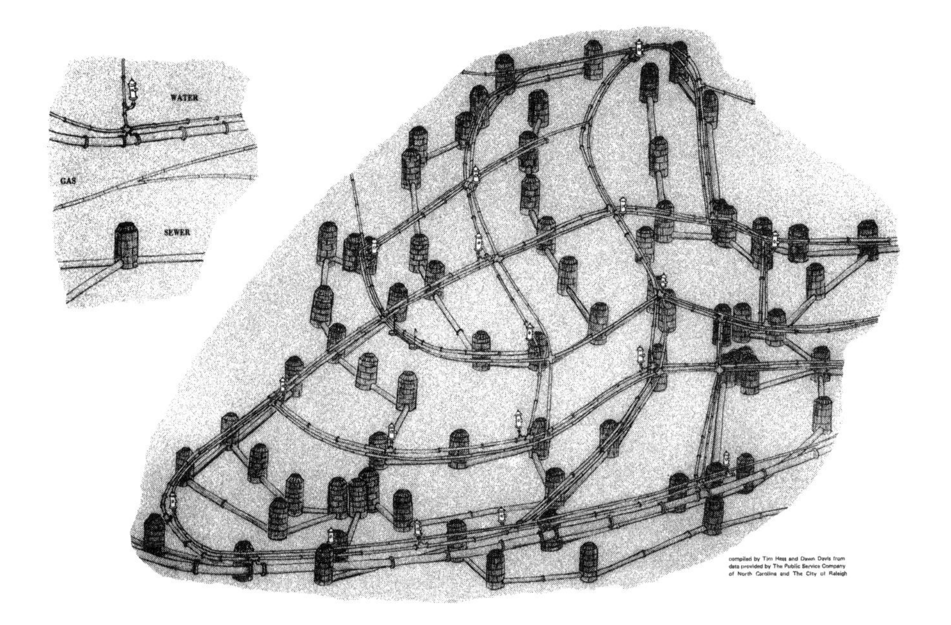

The social construction of this map—of the gas, water and sewer mains below the neighborhood in which Lot 126 is located—is hard to overlook, since, underground, it is impossible to *see*. (Drawn by Carter Crawford.)

ness to rely on the map is commensurate with our ability to suspend our disbelief in its veracity, but this amounts to a willingness to accept the map as an eye where the eye too no more than selectively brings into being a world that is socially construed.[32]

The temptation here always is to illustrate the truth of these assertions with outrageous examples. The effect is to protect the alibi by poking in it only . . . *the most obvious holes*. By parading egregious instances of map bias, the vast corpus that underwrites our daily living is allowed to evade inspection. A story in this morning's paper is classical. A local high school, Cardinal Gibbons, isn't where its address—and most maps—say it is:

> But the confusion about Gibbons doesn't stop there. Some maps show the school west of Avent Ferry Road, resting between Fraternity Court and Western Boulevard. "Every map that I've seen has us about a half a mile west of here," Kockx said. "No wonder we haven't grown. Nobody can find us."

The implication is that everything else on the maps is where it's supposed to be, that except for this bewildering—but explainable—error, maps really are windows on the world. *This* is the exception that proves the rule. When isolation won't serve, miraculous sleight of hand: our attention is turned to "propaganda maps" whereby the innocence of other maps is protected by blinding us to all but a small corpus of maps in which everyone can see—and happily acknowledge—the social construction of the image.[33] Or, a big deal is made about the failures of maps *in the past* to reflect the "real world." This leads to much self-righteous indignation over the loss of the learning during the dark ages when the lamp of learning was extinguished,[34] and endless froth over the placing of elephants for want of towns on the uninhabited downs of Jonathan Swift, thereby permitting *contemporary* maps to appear as the windows they have—presumably by dint of hard effort and the "scientific" attention to standards—triumphantly become[35]:

> The stations of this network are normally located 25 to 100 km (15 to 60 miles) apart and will have NAD 83 (North American Datum of 1983) horizontal positions, with differential positions accurate locally at the 1–3 cm level and absolute positions relative to the NAD 83 coordinate system accurate to the 5–10 cm level. Since GPS is three-dimensional, these stations also will have a vertical coordinate (ellipsoid height) associated with them. These ellipsoid heights can be converted to orthometric heights, the quantity obtained from leveling surveys, using geoid height information. NGSD currently publishes such geoid information from the high resolution geoid height model known as GEOID90. This geoid can provide 1 cm accuracy between points 10 km apart.[36]

Only by the slimmest margins does the map fail to be a window on the world, margins which, because we can control and understand them, no more interfere with our vision than does a sheet of window glass.

All you have to do is ignore the frame.

All you have to do is ignore the way the window isolates this view at the expense of another, is open at only this or that time of day, takes in only so much terrain, obligates us to see it under this light . . . or that. This is the sleight of hand: if you're paying attention to the glass, you're not paying attention to what you're seeing through the window. Not that accuracy is not worth achieving, but it was never really the issue, only the cover. It is not precision that is at stake, but precision with respect to what? What is the significance of getting the area of a state to a square millimeter when we can't count its population?[37] Who cares if we can fix the location of Trump's Taj Mahal with centimeter accuracy when what would be interesting would be the dollar value of the flows from the communities in which its profits originate? What is the point of worrying about the generalization of roads on a transportation map when what is required are bus routes? Each of these windows is socially selected, the view through them socially constrained no matter how transparent the glass, the accuracy not in doubt, just . . . *not an issue*.[38]

Look: here's Plate 86 in the *Times Atlas* with the Suez Canal running right up the gutter. Here's Israel and here's Jordan and running around through them in place of the usual international boundary line symbol is a string of purple dots and dashes: "Armistice Line 1949" and "Cease-Fire Line June 1967."[39] What is at stake here? Certainly it is not the *location* of the lines represented by these dots which everyone agrees . . . *are where they are*. What is at stake is not latitude and longitude, measured to whatever degree of fineness imaginable, but . . . *ownership*: this is what is being mapped here. This is what the fight is about. And the fighting was just as ferocious—maybe more so—before Harrison's chronometer beat its first second and long before we had Global Positioning Systems. With our total station we can get a satellite fix where we're standing at 31.31 N and 35.07 E; and whether we call it Hebron or Al Khalil, we will all agree that it's 31.31 N and 35.07 E. But because the map does not *map locations* so much as *create ownership at a location*, it is the ownership—or the ecotone or the piece of property or the population density or whatever else the map is bringing into being, whatever else it is making real—that is fought over, in this case, to the death.

Here, a second example, from the morning paper, *completely explicit*. The headline reads: "Raleigh neighbors don't want place on city's map." Here again the question is one of annexation, in this case to justify another annexation:

> What Raleigh really has its eye on is part of the lucrative Centennial Campus that stretches across 1000 acres next door. State law requires that annexed land have a certain number of residents—something the new campus doesn't have . . . The 340 people who live next door would fulfill those requirements as part of a package annexation deal.[40]

The objection is that because the city won't be maintaining the neighborhood's narrow, private streets, the residents will be paying for services they won't be getting, resulting in a kind of multiple taxation: "We're being taxed almost three times—by the county, by the city and by our homeowners' dues," complained one resident. Again, there is no question where any of these things are: city, county, subdivision, campus. All exist as property, thanks to the agency of maps, whose accuracy again is not in question, because maps do not so much record locations as connect them to a living. County, city? The role of the map, *which will be to establish this connection, to make it real in the lives of the residents* (and through their mutual enmeshment in the hierarchy of nested maps real as well in the lives of the rest of the city residents), will pass unobserved by all but the guy who wrote the headline.

Every Map Has an Author, a Subject, a Theme

"Mirror," "window," "objective," "accurate," "transparent," "neutral": all conspire to disguise the map as a . . . *reproduction* . . . of the world, disabling us from recognizing it for a social construction which, with other social constructions, brings that world into being out of the past and into our present. Preeminent among these disguises is the general reference map, the topographic survey sheet, the map, which without a point of view, gives us the world . . . *as it is*. Is any myth among cartographers more cherished than that of this map's dispassionate neutrality? So surely is this the north toward which cartographers point that they take its presence for granted, as though the neutrality of the general reference map were a fact of nature, a common truth like "all men are created equal" or "everyone's out for himself." Like these, its truth is little debated. It is just there, lodestone for a time of doubt. In most cartographic texts, the general reference map does a brief turn in the opening, where its existence, like that of the Virgin Birth, is blandly announced.[41] An undefined term, it then disappears, though like a palsied hand, its presence is sensed in every line. "We all know how a map works, right? Good. Then let's get down to business." It is like a cookbook: what does it matter what a cake *is*? Follow these instructions and you will be able to *make* one. Compile and scribe, proof and print: *that's* a general reference map. If you can hold it in your hands is there any need to discuss it? Or, the general reference map is brought on stage to clarify what something else *is not*. Fleetingly, like the conjurer's hat, it is spun through the discussion: magically, from its empty interior, materializes the rabbit of the thematic map.[42] This is claimed, in contradistinction to the general reference map, to have a subject or a theme. Or could I have that backwards? A map without a subject . . . would that be like a song without a melody?

What would a map be *of* that lacked a subject, unless the horror of the empty mirror? *Of* nothing, it would *be* nothing. It would not be. Unless it were to pop up in another universe, that of the mathematician perhaps, as an empty grid; or in that of the linguist as a crippled language, a grammar without words to embody it. A map is always *of* something, always has a subject, even when that something is a fiction alive exclusively in the map that is of it.[43] It refers out from itself to another map, to the world, to the Nature of which it is not. Of some*thing* (its

It is not just maps like this, from a manuscript in a library in 12th-century Turin, which embody their authors' prejudices, biases, partialities, art, curiosity, elegance, focus, care, attention, intelligence, and scholarship: *all* maps do.

subject), it is also through some*one* (its author), for its presence in the world is ever a function of the representing mind, and as such—it needs repeating—prey to all the liabilities (and assets) of human perception, cognition and behavior.[44] This is no more than to say that the map is *about* the world in a way that reveals, not the world—or not *just* the world—but also (and sometimes especially) the agency of the mapper. That is, maps, all maps, inevitably, unavoidably, necessarily embody their authors' prejudices, biases and partialities (not to mention the less frequently observed art, curiosity, elegance, focus, care, imagination, attention, intelligence and scholarship their makers' bring to their labor). There can be no description of the world not shackled (or freed—for this too is a matter of perspective) by these and other attributes of the describer. Even to point is always to point . . . *somewhere;* and this not only marks a place but makes it the subject of the particular attention that pointed there instead of . . . *somewhere else*. The one who points: author, mapmaker; the place pointed: subject, location; the particular attention: the aspect attended to, the theme—nothing more is involved (and nothing *less*) in any map. For example, a cartographer (the author, the one who points) maps the vegetation (the theme, the focus of attention) of Europe (the subject, the place pointed to).[45]

Seen this way, it is not that the general reference map lacks a theme, but that it has *too many*, or that they are too deeply interwoven, that the map is more subtle than simple, too complex to bare in a single word—which words therefore are dispensed with altogether, as great novels today get along without the subtitles that adumbrated the themes of earlier ones, *Candide ou L'Optimisme*, *Emile ou L'Education*. Thus, not *Europe or the Vegetation*, *Transportation*, *Topography*, *National Boundaries*, *Cities and Points of Interest*, but simply (and more grandly) *Europe*, as we say *Ulysses* or *Love in the Time of Cholera*, with respect to which, simply because they are not itemized, we do not assume any lack of "themes." Perhaps the issue is essentially one of euphony, that on first hearing, "Vegetation Map" *sounds* reasonable, whereas "Vegetation–Physical–Political–Urban Map" *sounds* silly and cumbersome. Whereas it is a form of snobbery to prefer the seemingly elegant ("Vegetation Map") to the merely utilitarian ("Vegetation–Physical–Political–Urban Map"), it is a form of madness to confuse the titles with the content, and so come to mistake the "Vegetation Map" for a map of vegetation, or the map of "Europe" (elegant cover for "Vegetation–Physical–Political–Urban Map") for a map of Europe. The former is to mistake the theme for the subject; the latter to take the map for the subject itself, as though it were possible to have a map purely of its subject, of Europe, not of the *vegetation* of Europe, or the *topography* of Europe, or the *cities* of Europe *today*, but, you know, of *Europe itself*, as it is, once and forever, warts and all.

But sooner this hallucination than a cacophonous title, even if such self-deception should result in the articulation of a class of maps founded, not on content, but on names: those called by their themes (vegetation, urban, climate) and presumptively partial (*the thematic map*); and those named after their subjects (Europe, North America) and presumptively impartial (*the general reference map*). That such nominal classification bears but the slightest relationship to the subject of its attention (the maps) is but a trivial sign of panic, sad, but innocuous. The poison lurks in the ascription to the maps named after their subjects (the general reference maps) of, initially, literal impartiality; that is, not being partial (as thematic maps are), to either vegetation *or* national boundaries *or* topography; and since not partial in this way, literally *im*partial (that is, *comprehensive*, as general reference maps are supposed to be). Soon, however, impartial, ceases being heard as not partial but comes to be heard, figuratively, as impartial; that is, as fair, free from bias, disinterested; as in John Dewey's "impartiality of the scientific spirit," that is, as objective, dispassionate, even neutral; until ultimately purely and totally of the subject, without mediation, *transparent*.

Cartographers talk as if this were all well understood. The editors of *Goode's World Atlas*, to exemplify, are nicely outspoken. As they write in their introduction,"Because a well-drawn map creates an aura of truth and exactness, the cartographer should caution the reader against interpreting the generalized data too literally,"[46] but frequently they do not mean what they say, they rarely practice what they preach, and have managed to order their maps so as to preserve the implications of transparency for the general reference section. Most of all they are handicapped by the ferocious power of the maps to speak for themselves. The effect is to have created an artifact that says one thing wrapped in words that claim it is something else.

To illustrate: in their introduction to the "regional section," these editors write of their "environment maps" that their boundaries "as on all maps are never absolute but mark the center of transitional zones between categories."[47] One wants to applaud: *wonderful sentiment*. But that's all it is, a sentiment. For certainly it is not true, as stated, in the general case (unless we are to exorcize maps of their cadastral and political content), and is adhered to in no other, for where the idea of the *zone* has merit, there is invariably a fine black line (as that separating Mediterranean agriculture—in a stippled yellow-green—from deciduous forest—in tan); and where the idea of the *line* has merit there is invariably a zone—depending on the scale, up to 20 miles thick—engulfing a very broken line (as between Germany and France). It is a kind of nominalism which, having insisted that a boundary is not a line, feels perfectly free to draw it as nothing else.

Suspended Between Faith and Doubt

What is its cost? At public meetings citizens peer at small-scale maps on which city planners have scrawled road proposals in markers wide enough to be seen from the back of the council chamber and then, during a break, have heart attacks when they go up close to find the road on top of their homes. Reassured about their homes by careful explanations of the road's actual width, they nevertheless continue to accept the inevitably and accuracy of the rest of the map . . . including the proposed road. Why shouldn't they? Doesn't the map merely . . . *reproduce reality?* If it does, everything on the map is real. If it doesn't, nothing is. Not only would the proposed road be open to debate, but so would the course of *this* stream and *that* political boundary. But if they *are* real, then—except for unwitting error, an unintended failure of accuracy—*everything* on the map is above discussion. This *is* where the stream runs and that *is* where the boundary lies, and that *is* where the road will be built. Can we have it both ways? *We have to.* For the map to enable past or future to become part of our living *now*, it has to be able to connect it to a *here*. Otherwise, paralyzed by doubt, we are reduced to inaction: "Well, we want to plant a hedge, but until we really know where our property line is . . ." Yet unless we continuously *question* the map, doubt—yes—its accuracy, but more critically what of past or future it is linking up to the present and how it is doing so, the map will *disable* us from acting with intelligence and grace, will doom us to a living that is fatally flawed, partial, incomplete: "Well, we planted a hedge there, but none of the maps we looked at showed the city's plan to widen the road." Between doubt and conviction we must perpetually cycle: "We forfeit the whole value of a map if we forget that it is *not* the landscape itself or anything remotely like an exhaustive description of it. If we do forget, we grow rigid as a robot obeying a computer program; we lose the intelligent plasticity and intuitive judgment that every wayfarer must preserve."[48] At once the map *is* and *is not* the terrain:

> "The map is not the terrain," the skinny black man said.
>
> "Oh, yes, it is," Valerie said. With her right hand she tapped the map on the attaché case on her lap, while waving with her left at the hilly green unpopulated countryside bucketing by: "*This* map is *that* terrain."
>
> "It is a quote," the skinny black man said, steering almost around a pothole. "It means, there are always differences between reality and the descriptions of reality."
>
> "Nevertheless," Valerie said, holding on amid bumps, "we should have turned left back there."
>
> "What your map does not show," the skinny black man told her, "is

that the floods in December washed away a part of the road. I see the floods didn't affect your map."[49]

But the floods didn't wash everything away, they were not those that only Noah survived. Poised, suspended, between faith and doubt, we must make our way through the world of maps.

CHAPTER TWO

Maps Are Embedded in a History They Help Construct

Can the truth really be so hard to find? It all depends on where you're standing. Every view is taken . . . *from somewhere,* every view is but *one perspective on the common scene.* The variety this implies is bewildering (or beguiling—this too is a matter of perspective), but it is also less than at first it seems. The view from this bench in Raleigh, North Carolina *is* privileged, but not especially. If I move to its other end, I lose a little of the shade and everything has shifted . . . but not much. I will still be able to watch the squirrel chasing its tail and the bicycle will not cease leaning against the tree. The sky will be as blue . . . *even from another bench.* The view here is from America at the end of the 20th century. It is not that from England in the 16th century or China at the height of the Dong or Egypt during the 18th Dynasty. It is not that from the back of Red Cloud's horse before the Fetterman Fight or that of the Tellem on the Bandiagara escarpment before the coming of the Dogon or that of the man whose handprint can still be seen on the wall of this cave in the Pyrenees. It is not a view from a satellite or the moon, Mars or Alpha Centauri. It is not that of God.

It is mine . . . *wherever it is from* . . . but this too implies more freedom than I well can claim. It is not to be a determinist to acknowledge the claims of parents and birthplace, the demands of routine, to admit—even—to a certain rut that it is less than easy to get out of. It is to admit the course of growth, the sway of development, the power of history.

Growth, Development, History

Randall and Chandler, my two fine boys, and I have lived together since before they were born. For 17 years I have supported their growth and participated in their development, helping them turn from mewling, all-but-helpless infants incapable of controlling their sphincters into the assertive and all but autonomous hulks who last summer roamed on their own around Manhattan. I think, though, that they have always felt like this, capable, that is, of purposeful action. *They* have always felt more or less powerful, more or less autonomous . . . or at least no less so than they do now. There was never a time when they *felt* like the babies or toddlers or little kids they appeared to be from my perspective. "Baby" and "toddler" and "little kid," after all, are adult words for children who, however small and muscularly undeveloped, never (or hardly ever) say of themselves, "See how weak we are," but always (or usually), "Look at us! See how strong we are!"—as though they were Joe Weider master blasters pumped for a Mr. Olympia contest instead of the 60-pound weaklings they inevitably are. Tom Watterson plays in the gulf between these perceptions. In one of his *Calvin and Hobbes* strips Susie asks Calvin if she can play with him and his tiger. "Hobbes and I are not playing," Calvin archly informs her. "We're doing important things, and we don't need you to mess them up."[1]

All this is exactly how I felt about things when I was growing up—that is, I was competent, I was strong, what I was doing mattered, was important. I don't think I was that different from other kids I knew, but I was sufficiently secure in my feelings to send off my idea for a rocket to Charlie Wilson, then Eisenhower's Secretary of Defense. And the response I received from the Deputy Director for Special Activities did nothing to diminish my sense of being, at age nine . . . *on the cutting edge*. But I *know* I am different today. Looking back I see that I can do things now that I could not do then, however grownup I may then have believed myself to be.[2] I have more practice at thinking. I can reverse operations and start them in the middle to work my way out in either direction. I have a bigger vocabulary, and I can make more subtle discriminations. I can get into movies that once I couldn't. Because I have a job, I can even pay my way. When I walk into a pornographic bookstore, no one tries to stop me. There are a lot of things I can't do too. I can't sit on my mother's lap the way I used to, or fit into the clothes I wore when I was nine. I can't play with toys the way once I did, insinuating myself unself-consciously into the very cab of the little truck that once filled my hand. And I can't ever feel the way I did in the days before the time I hit my wife when I was drunk.

If I try to disentangle the threads twisted together in this braid of my experience, I can easily grab hold of three. The most obvious is simple physical growth: I weigh 150 pounds more than I did when I was born and stand 4 feet taller than I did then. But I'm not just bigger: I'm better integrated. I can do things that require the subordination of one part of my body to another, that force me to differentiate short-term lusts from long-term needs. Not only can I ride a bike and dance and do aikido, but I can type and speak and write, as now, in complicated sentences. So the second thread is that of development, that is, my increased differentiation, articulation and hierarchic integration. This development did not (and does not continue to) occur *in vacuo* but in the United States in the 'Fifties and 'Sixties (and 'Eighties and 'Nineties) when certain things were (and now other things are) possible and certain things weren't (and still aren't). No matter how often I dressed up in what I wanted to believe was the costume of a medieval knight, I could never ride off to King Arthur's court. That was an historical possibility . . . *foreclosed*, precisely as Russia's Sputnik *opened* the way to the scholarship monies that permitted me to attend graduate school. And obviously the third thread is history, the way my growth and development was (and continues to be) shaped by the ceaselessly changing social and physical environments that I at the same time *collaborate* on bringing into being.

These changes in me and my kids constitute the central reality of my experience, and I see these three faces of the unfolding we call life at every scale.[3] Systems, processes, things of every kind seem to get bigger or smaller, to grow more or less hierarchically integrated, to interact with other things engaged in similar processes to make our history.[4] It is these that we see taking place at the scale of atoms and molecules in the stories we currently tell about the early history of the universe.[5] It is these that we see unfolding at the scale of biological organisms in the story we call evolution.[6] It is these that we see occurring in colleges and corporations, in families and cities, in national governments.[7] In each domain I cannot help seeing the same three threads of growth and decay, of development and pathogenesis, of history.

Maps Themselves Don't Grow (or Develop)

Though it would be silly to ignore the way maps come into being and subsequently disappear, I do not wish to claim that the map artifacts themselves grow or develop, although Christopher Tolkien has documented just such a process in his father's construction of a map of Middle Earth:

> It consists of a number of pages glued together and on to backing sheets, with a substantial new section of the map glued over an earlier part, and small new sections on top of that. The glue that my father used to stick down the large new portion was strong, and the sheets cannot be separated; moreover through constant folding the paper has cracked and broken apart along the folds, which are distinct from the actual joins of

the map sections. It was thus difficult to work out how the whole was built up.[8]

Here we see not just growth and decay, but also development, for what J. R. R. Tolkien did was to continuously differentiate, articulate and hierarchically subordinate the parts of the Middle Earth he was creating . . . *interactively* . . . with this map; so that history too appears here, in the way the map takes as given certain aspects of Middle Earth previously worked out, even as it—precisely—generates others. Old Sanborn maps grew like this too, layer upon pasted layer, as the cities they mapped changed, as they grew and developed, the maps interacting with the insurance and firefighting systems of the cities they represented to help bring forth the history they would in time come to embody.[9] The stick charts of the Marshall and Caroline Islanders also grow this way, literally get larger, coconut-palm rib by cowrie shell, and stick by stone.[10] The ephemeral maps of the Inuit, scratched in the dirt, traced in sand and

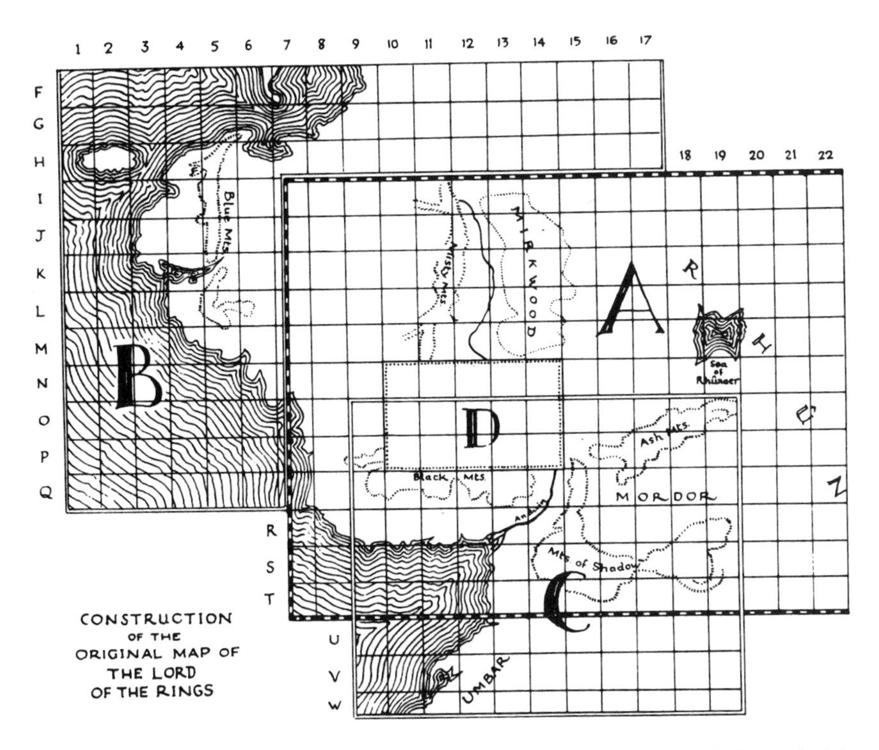

This is a diagram of J.R.R. Tolkien's construction of the original map of *The Lord of the Rings*: first he drew one part, then he superimposed another, then extended a third, and so on. (From J. R. R. and Christopher Tolkien, *The History of Middle Earth, Volume VII*. Copyright 1989 by Frank Richard Williamson and Christopher Reuel Tolkien. Reprinted by permission of the Houghton Mifflin Company, Inc., and HarperCollinsPublishers.)

snow or allowed to evaporate in the air, also grow this way,[11] to say nothing of the sketch maps *we* casually make, mark by mark.[12]

And yet there's an important sense in which all of these undoubted maps are quite marginal to what we mean when we talk about maps in an unself-conscious way. The reference then is to printed maps, typically produced in enormous numbers on high-speed offset presses. Except at the edge where the ink is being laid onto the paper at a hundred miles an hour, these maps don't *grow* either, at least not in the way we ordinarily use that word. They are slapped out—shh! shh! shh!—onto the paper elevator at the end of the press, and except as they fall apart from constant use or are chewed up by the dog or rot (or fail to rot) in a landfill, they don't decay either. Nor do they much develop. We might scrawl a note or a route or a destination on a map, and so increase its level of differentiation, but this is not often the case and usually the map artifact itself neither grows nor develops.

But Mapping and Mapmaking Do

What does grow and develop, however, are the systems or processes or things we refer to when we say "mapping" or "mapmaking." These words do not mean the same thing. "Mapping," as Robert Rundstrom has pointed out "is fundamental to the process of lending order to the world."[13] What he is speaking of here is the way we humans make and deploy mental maps. Maybe 30 years ago the unqualified assertion that humans created and used mental maps could have been greeted with caution (if not downright skepticism), but not in an age when it is possible to assert without being in any way provocative that *bees* make and use mental maps.[14] Remarks being made today by biologists eerily echo those made 20 years ago by psychologists. Where in 1969 the psychologist David Stea, pondering the geometry of mental maps in humans, assumed that "all persons form conceptions of those significant environments too large to be perceived, i.e., apprehended, at once,"[15] in 1989 the biologist Talbot Waterman, pondering the geometry of mental maps in animals, observed that "whatever its modality the basic geometry of animal maps is a matter of great interest but little certainty."[16] What is remarkable here is the absolutely taken-for-granted quality of the animal maps in question. Given the wide-spread assumption today that animals make maps,[17] it is hard to imagine that adult humans don't; and evidently, humans and their immediate predecessors have used mental maps for millions of years, an ability selected for by their self-evident utility to an increasingly mobile genus.[18] That is, the growth, development and history of the *mental* map are questions of *evolution*, the gradual appearance of the trait taking place over the many generations it took anatomically modern *Homo sapiens* to evolve.[19] At the same time it is an ability that flowers *in us* now—today—as we grow, develop and interact with the world in our modulation from fertilized egg to adult. Whether ontogeny recapitulates phylogeny may be moot,[20] but the universal ability to make and deploy mental maps in all human populations is not.[21]

There is, therefore, no doubt of the mapping abilities of those we still term "primitive." It is time to acknowledge that people like Catherine Delano Smith and Malcom Lewis are simply wrong when they speak of human groups with cognitive abilities less than ours. Relationships among *spatial cognition*, the *ability to make* maps, and their actual *production* are not straightforward, and the failure of the latter cannot be taken to indicate an absence of the former. Anyone who has tried to collect so-called "mental sketch maps" from college students knows how often maps exhibiting no more than "topological relations" are collected from individuals who have manifestly mastered "formal operations." Everywhere we find examples of those whose behavior in this or that circumscribed domain is the same as that exhibited in much earlier developmental stages than the one achieved and exhibited globally. It is simply not possible to assess general levels of intellectual development from the "sophistication" of this or that isolated gesture. This is not only because we develop abilities over different contents and in different domains at different rates, but because *we enter each new content area and domain in some sense as if each time we were starting again from scratch.* We then proceed, microgenetically, as fully operational adults, to pass through sensorimotor, preoperational, concrete and formal operational stages—to refer only to Piaget's typology.[22]

How disturbing, then, to read in Lewis' "The Origins of Cartography" of "cultures in which cognitive development, even in adults, terminated at the preoperational stage."[23] This would mean, for every content domain, that these adults could repeat but not reverse operations (for example, they wouldn't be able to reverse a route to return home), would fail to justify assumptions (even in heated debate), would find it difficult to decenter from a given aspect of a situation (that is, to take another's point of view, including those of gods or animals in rituals and celebrations), and would be unable to coordinate perspectives (which is to say they wouldn't be able to create an "areal view") among other limitations. Such adults, in other words, would be behaviorally indistinguishable from, say, your 5-year old, and therefore (presumably) incapable of producing anything we might recognize as a map. Bluntly put, no such culture of *Homo sapiens* is ever known to have existed.

But if the cognitive attainments of individuals are invariant across culture, what is it that is "primitive" about "primitives?" Very little probably. Certainly the use of "primitive" which was widespread to describe non-Europeans when the history of cartography was struggling into existence was unjustified (as was the similar characterization of medieval mapmakers[24]). Such judgments concerning the sophisticated worlds of the Dogon and Hopi can only be explained as ignorance fueled by chauvinism (a behavior evidently preoperational in its inability to decenter from the labeling social group). Yet rejection of the pejorative implications of "primitive" cannot be allowed to mask the reality that differences among groups exist. Just because a squirrel can map its environment doesn't mean it can communicate this knowledge to others. Just because bees can both map and communicate such knowledge to other bees (through their notorious waggle dance), doesn't mean they can make maps, that is, produce the artifacts we unhesitatingly accept as such. And just because humans *can* make maps, doesn't mean that they do, at least as a matter of course, in their everyday taken-for-granted world.[25] Although "development" may seem appropriate only for describing systems that change over time, the term, as used by physicists, biologists and psychologists (notoriously Heinz Werner,[26] but also Piaget[27]) characterizes the *degree of organization* of any system. In this way it may be used to compare different co-existing systems, and I will be using "development" in this way to compare the degree of organization of the mapmaking systems of different societies. Yet, it is the sense of *transformation*, from being unable to map the world, to being able to, to being able to communicate it to others, to being able to produce artifactual maps, to living map-immersed in the world that I am most thinking of when I speak of the growth, development and history of mapping and mapmaking.

To Live Map-Immersed in the World

What exactly do I mean when I refer to being map-immersed in the world? I mean being so surrounded by and so readily and frequently consulting and producing maps as not to see them as different from the food that is brought to the table or the roof that is overhead or the culture in general that is apparently reproduced . . . *without effort*. Three years ago I tried to understand what this might mean by collecting every map my family encountered, used or produced in its daily life. Intending to keep this up for 30 days, I gave up after 20 so numerous were the maps involved.

On the second day into the period my then 14-year-old son, Randall, produced two elaborate maps of "Rebel Installation SR 543-k3" for the role-playing universe he was then running for a group of friends; during the period in question he was obsessively involved with these maps. My son Chandler, then 12 years old, made two maps during the

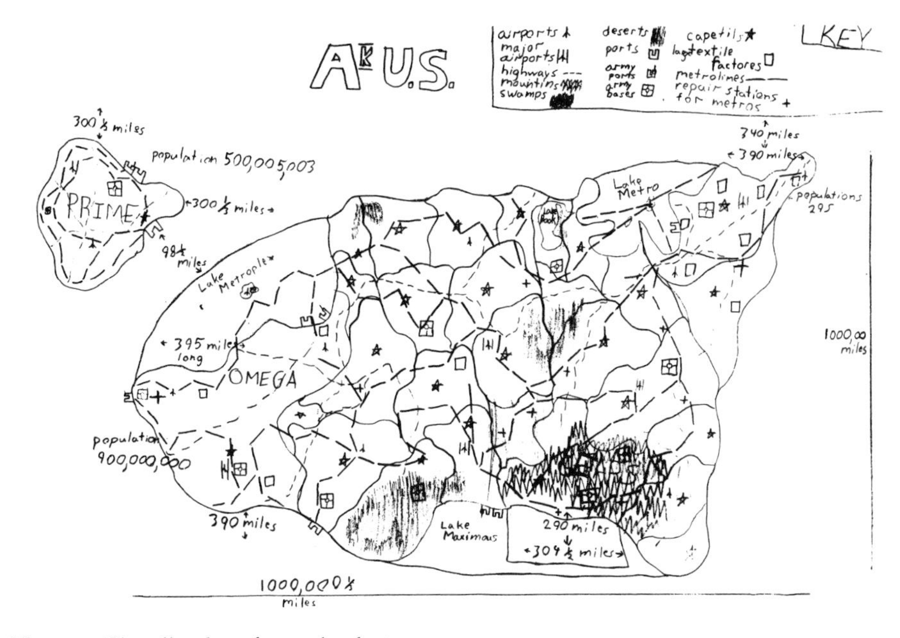

The map Chandler drew for a role-playing game.

period for a school project on France. One was of departments, capitals and major rivers; the other was for a tourist brochure of attractions along the Seine ("France: The Country of Romance"). He also spent a lot of time during this period drawing elaborate plans for water parks (he drew as many as four or five a day); produced a map for a role-playing scenario; and quite spontaneously created a map of the world, apparently stimulated by a visit from Tom Saarinen who projected on our dining room wall slides showing maps from his National Geographic Society study of world views.[28] During this period nine maps were drawn for Pictionary games in attempts to evoke "Brazil," "Taiwan," "Los Angeles," "Illinois," "East Coast," "trip," "map," "area code" and "foreigner." Maps were used in the game Risk and showed up on packaging, in advertising and as editorial content in newspapers and magazines. Maps played central roles in numerous social situations. On the first day I gave my wife Ingrid maps of bus routes I had collected for her in Spokane and Portland to use in her capacity as a member of the Raleigh Transit Authority. On day two, the two of us consulted a pair of Amtrak maps to plan our summer train trip. On the third day I found my older son with *Volume IV of the Mid-Century Edition of The Times Atlas of the World*. "What's up?" I asked him. "Do you think you could Xerox this? I need it for my report on the Canary Islands." Two days later, Ingrid took

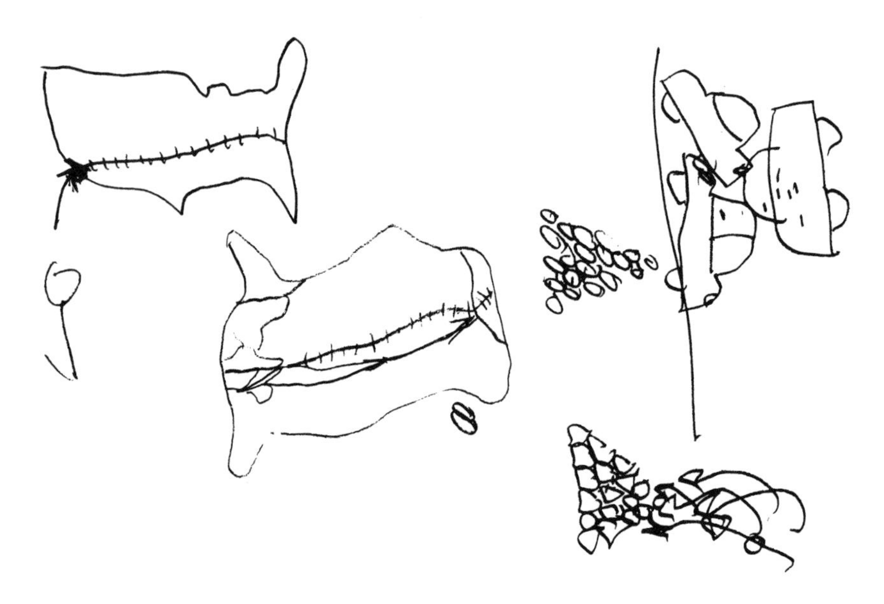

These are maps sketched during a game of *Pictionary*.

our *Goode's* off the shelf to show Chandler the route of the trip we'd planned for the summer. A day after that Randall and his friend, Garland, used a city road map to clarify the bike route we'd taken to see *Beverly Hills Cop II*. This led to a discussion of distances in which Garland used the map's index to find Walden Pond Road, and then he calculated the distance he'd biked to get there. Five days later Randall took a city road map with him on his Sunday bike ride to Wake Forest. Five days later still, on a bus trip with my father to pick up some replacement speakers, we conferred about our route while consulting the map on a bus stop kiosk. On the way home, when he observed that we were following a different route, we looked at the map on the bus schedule. Talking on the phone 2 days later, we each consulted our own copy of a city map as we tried to find various locations pertinent to our discussion. Two days later, a friend came by to deliver a pair of maps we needed to have mounted on foam core for a presentation to the Raleigh City Council. On our way to the stationers, we delivered to another friend a yard sign promoting our cause. The yard sign had a road map on it. That night at dinner, Chandler asked about Greenland on the Surrealist map of the world on a T-shirt I was wearing promoting R.E.M.'s *Little America* album. On his shirt, over the left breast, was a logo constructed around the outline of North Carolina.

How different all this is from the experiences I have had in Zinacantan, a community of Tzotzil-speaking native Americans in southern Mexico's Chiapas highlands. In the many days I passed there in the

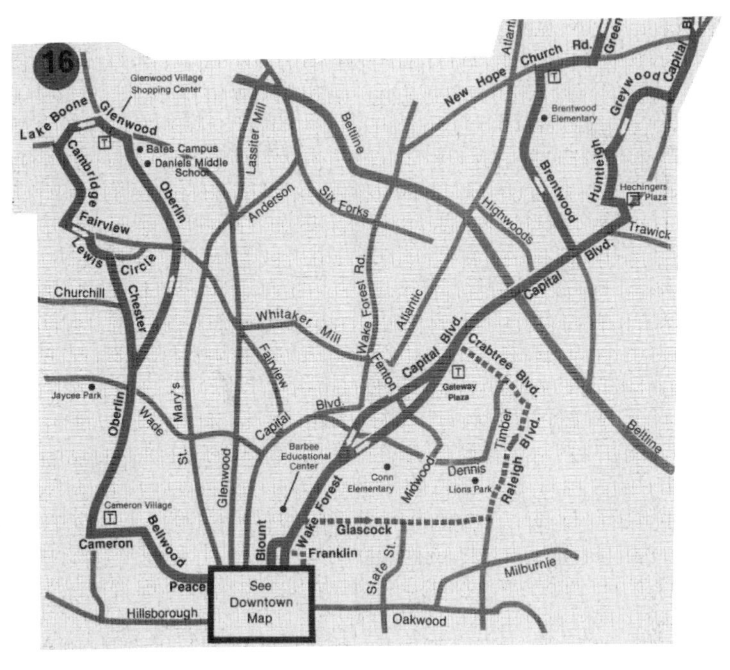

The map my father and I consulted on our bus trip: in a map-immersed society there is no end to these entirely ordinary maps.

home of my friends, I recall seeing but a single map, in the textbook of one of the older boys who was studying Spanish in school. It was something I took pains to see, being curious about what he was learning in class. Maps appeared nowhere else in their home, unless perhaps in the logo of a Mexican government agency crudely stenciled on the burlap of the bags used to store corn. Maps were not drawn in the context of games or in the talk about the community that flowed endlessly around the fire. Kids did not discuss their day with their heads over a map. Fathers and sons did not trace out bus routes. Mothers and daughters did not turn to the atlas to work out a summer vacation trip. There were no books in the home, no magazines or newspapers. The *mental* atlas was continuously consulted. Geographic names peppered every discussion. Detailed knowledge about the twists and turns of paths was taken for granted. The layout not only of the town they lived in, but also of the large nearby Mexican center of San Cristobal, was frequently referred to as indeed was the geography of the State of Chiapas. At the time when there was much talk about our first landing on the moon, I witnessed my friend's father use a cup of coffee and his fingers to describe to *his* mother how on its return the capsule would splash into the ocean and be rescued by a ship. I did not get the feeling that the explanation made much sense to her for whom the moon and the Virgin remained fused in a syncretic amalgam worked out in the centuries following the Conquest. It was no lack of cognitive ability that interceded, but rather a lack of knowledge about the world of NASA and Apollo and the astronauts that we absorbed from *Life* magazine and television and through 12 more or less mandatory years of schooling. Certainly my friend or his peers drew and helped to construct and interpret the maps and airphotos introduced into their lives by the anthropologists of the Harvard Chiapas Project,[29] but maps were not a deep part of their living. You see photos of them with government functionaries peering at maps at the founding of an *ejido*, and we know that their ancestors 20 and more generations ago produced the *lienzos* that even today are used in the courts to adjudicate land disputes, but maps do not play the role in their lives that they do in mine. Maps remain special, rare, precious.

Some Societies Are Bigger Than Others

What is the difference between my Zinacanteco friends and me, between their world and mine? If I follow my first thread, it is simply that there are more maps in my world than in theirs. I have no idea how many maps I have in my house, but the number is enormous. Even in homes less involved with maps than mine the number is high, even if they're only the ones in the phone book. Most maps may be crammed into glove compartments or kitchen drawers, but it is precisely that casual taken-for-granted quality that is the point. And most of these maps exist in numbers of copies running into the hundreds of thousands. The world these maps encode is much larger, too. Raleigh, the not terribly large city in which I live and in whose political life I am deeply involved, has more people living within its boundaries than there are Tzotziles altogether, that is, than there are Zinacantecos and Chamulas and Pedranos and all the rest of them, each with their own "center," their own patron saint, their own . . . *world*. And whereas these "centers" are not integrated into a larger Tzotzil world, and scarcely into a Mexican one, mine is *self-consciously* knit into many larger overlapping ones. As a matter of course I traveled thousands of miles to be able to stand before a group of geographers and for 20 minutes read an earlier version of this chapter in a room I needed a plan to find in a city I needed a map to even begin to understand.[30]

The greater size of my world, the greater number of persons integrated into it, has two implications. In the first place, maps are required for us all to keep track of each other and what we're up to. They manage this by connecting us *through them* to all the other aspects of the vast system of codes, laws, contracts, treaties, covenants, deals *and so on* in which we have immersed ourselves. But in the second place, maps *allow* us to keep track of each other: *the specialization required demands a population of the size it permits to function*. I imagine there is some threshold above which mapmaking emerges in a society, and below which it doesn't, not because its members are incapable of making maps, but because the society is too small, with too little specialization to either require or support it. What is this population? It would be interesting to try and figure out, but I am hypothesizing that it will be rather large, much larger, for example, than traditional Micronesian or Inuit groups, even taken as a whole. The wonderful abilities of the great Micronesian pilots confirm rather than undermine this thesis. Theirs is the navigational skill of a Mississippi tow boat pilot, and the two groups of pilots are treated with an equivalent extravagance of position and praise. The Micronesian pilots and teachers undoubtedly make their well-known charts (though increasingly these are made by others for sale as curios), but there is little other evidence in their society of mapmaking and using (until recently, that is, until its integration into ours, into the world society of post-Fordian capitalism). In the society of the tow boat pilot, on the other hand, other kinds of charts are used by other kinds of pilots to sail the oceans, fly the skies and ply the roadways; nor is anyone surprised by an individual who makes his living sailing, flies for fun and uses a road map to get around his homeport. Besides these direct navigational aids, such pilots might be expected to consult weather maps, bathymetric diagrams, charts of rivers and ports, plans of his ship. They are *immersed* in a world of maps and charts and plans in a way their Inuit, Aboriginal and Micronesian counterparts are not. And this *is* related to the simple size of the society of which they are parts, for tiny societies cannot differentiate themselves to the degree that larger ones can.

Some Societies Are More Developed

Clearly, however, sheer size would be more a liability than an asset were the population not differentiated, specialized, hierarchically integrated, and indeed were these conditions not met, it could be doubted that a society could grow so large. Certainly there is labor specialization among the Zinacantecan—the women herd, the men hoe, and there are shamen, musicians and others—but the greater part are farmers, terrific generalists. There are no air conditioner repairmen among the Zinacantan, no lawyers whose practice is limited to problems with pension funds, and certainly there are no surveyors, cartographers, map engravers, copy camera operators, plate makers, pressmen, or sales representatives for commercial producers of maps. It is the development of this system of production with the technology it implies of generation, manufacture and distribution that in the end most radically differentiates mapmaking cultures from those that aren't.

This specialization penetrates our consciousness and thereby differentiates us not only from each other, but as a society from those societies whose consciousness remains more whole. Less specialized, such societies are . . . *less alienated*, less caught up in the logics of mapmaking, in the logics of print, in the logics of reproduction with their attendant demands for continuity and uniformity, abstraction and quantification. They are less caught up in *all* the logics required to integrate such highly differentiated masses as our society and others like it have become.[31] The maps made by members of less alienated societies are different from ours. Often the *process* is more important than the *product*. Rundstrom has written:

> During field work in 1989, one Inuk elder told me that he had drawn detailed maps of Hiquligjuaq from memory, but he smiled and said that long ago he had thrown them away. It was the act of making them that was important, the recapitulation of environmental features, not the material objects themselves.[32]

Others have stressed other differences. Harley has quoted William Cronon to the effect that:

> . . . even the objectives of English and Indian naming of landscape features were different. Thus, the English "frequently created arbitrary place names which either recalled localities in their homeland or gave a place the name of its owner" while the "Indians used ecological labels to describe how the land could be used."[33]

David Turnbull pulls on an altogether different thread, one related to cultural variations in indexicality. In his *Maps Are Territories*, Turnbull compares three Aboriginal-Australian maps with a sheet from the British Ordnance Survey, *interpreting* the former and *interpellating* the latter. (An interpellation is technically a formal bringing into question—as in a European legislature—of a ministerial policy or action, but this is what the 43 questions Turnbull puts to the Survey sheet amounts to.) This has the effect of "deconstructing" the survey sheet—bringing to the surface its hidden assumptions (hidden because taken for granted, because transparent to our sight)—at the same time that the interpretation of the *dhulan* brings to the surface their hidden assumptions (hidden from us because hermetic, because unshared):

> Aboriginal maps can only be properly read or understood by the initiated, since some of the information they contain is secret. This secrecy concerns the ways in which the map is linked to the whole body

> of knowledge that constitutes Aboriginal culture. For Aborigines, the acquisition of knowledge is a slow ritualized process of becoming initiated in the power–knowledge network, essentially a process open only to those who have passed through the earlier stages. By contrast, the Western knowledge system has the appearance of being open to all, in that nothing is secret. Hence all the objects on the map are located with respect to an absolute co-ordinate system supposedly outside the limits of our culture.
>
> One could argue that in Western society knowledge gains its power through denying, or rendering transparent, the inherent indexicality of all statements or knowledge claims. In the Western tradition the way to imbue a claim with authority is to attempt to eradicate all signs of its local, contingent, social and individual production. Australian Aborigines on the other hand ensure that their knowledge claims carry authority by so emphasizing their indexicality that only the initiated can go beyond the surface appearance of local contingency.
>
> In the light of these considerations we should perhaps recognize that all maps, and indeed all representations, can be related to experience and instead of rating them in terms of accuracy or scientificity we should consider only their "workability"—how successful they are in achieving the aims for which the were drawn—and what is their range of application.[34]

Though in *Maps Are Territories* this does segue into a comparison with Western maps from the Ptolemaic and medieval traditions, Turnbull's is not the unbridled relativist posture it might at first appear. What it does emphasize, though, is the way we have imposed on the study of world cartography not only criteria of our own—that is, generated from within, from within our own culture—but among these selected for special emphasis the very ones we most labor to produce in our own work (any history written under this aegis will inevitably construe our cartography as the acme of perfection). Accuracy, to recur to this issue, is not a measure that stands *outside our culture* by which other cultures may be evaluated but, rather, is a concept from *within our own* culture that may be irrelevant in another. In yet a third culture, accuracy may be an issue, but with respect to what? Certainly our topographic surveys do not, as the *dhulan* do, accurately represent the "footprints of the Ancestors." Of course if you don't believe in the Ancestors, then it's all a bunch of primitive nonsense anyway[35]; but such cultural absolutism is not only repugnant (isn't this precisely what we condemned in Iran's Khomeini?) but impossible to justify.

But when all this is said and done, doesn't it remain true that societies with high degrees of labor specialization are more *advanced* than those that aren't? Doubtless this choice of word is not simply wrong but subtly and probably intentionally misleading. Yet such societies *are* more developed, if by that we mean that they are more differentiated and hierarchically integrated. In addition to simple growth in the size of the society and the sheer number of its maps, the *varieties* of maps—and the *relationships* among them and to the society that generates and uses them—have grown increasingly differentiated, have become increasingly well articulated, have found themselves increasingly hierarchically integrated.[36] Furthermore the mapmaking and distribution system is increasing in each of these ways even as I type this. That is, not only is our society more develop*ed* with respect to mapmaking than that of the Aborigines and the Tzotzil, it is develop*ing* at a more rapid clip in that it is differentiating, articulating and reintegrating itself in this domain more rapidly. Stephen Hall's *Mapping the Next Millennium* constitutes a sketch of this wavefront.[37]

But because of this are we better off in any deep way than the Tzotzil? Are we happier? Are we more satisfied? Do our days pass with greater intelligence? Hard to say, though what is easy to say is that the Tzotzil are in our orbit as we are not in theirs. The mapping impulse was deeply implicated in the Spanish Conquest, and the hold of the landlords over the subsequent centuries was insured and perpetuated in maps. This is to say that it was the differential development of mapmaking no less than other differences between the Spanish and Americans that resulted in the domination of the latter by the former. Thus are we catapulted into history.

Our Histories—Entwined—Are Different

Evidently the making of artifactual maps *originates* in many impulses, even as writing does, but neither seem to *develop* in the absence of a need to keep records.[38] The reasons for this were undoubtedly manifold, and while relevant, too hypothetical to reasonably treat here. Denise Schmandt-Besserat's hypothesis for Mesopotamia involves the necessity of accounting in long-distance trade[39]; Mary Elizabeth Smith's hypothesis for the Mixtec invokes the complexities of land ownership amidst dynastic turmoil.[40] There are other hypotheses, but almost all of them assume that what was at stake was *control* of social processes in rapidly expanding groups. A variety of modes ranging from the linguistic through the logographic to the purely pictorial—and including mixtures of each—were used to record qualitative and quantitative information in both spatial and temporal dimensions. Signs that originally developed as names in narrative descriptions of lineages or routes were adapted as pictures on maps—and *vice versa*. Over time, in accordance with structuralist principles, the notation systems differentiated: *temporarily* ordered information (such as lineages and routes), which was recorded using logographic and *linguistic means*, developed into what we recognize as writing (toward history and descriptive itineraries); whereas *spatially* ordered information (such as land ownership, the number of sheep in various fields belonging to different owners, and routes), which was recorded using logographic and *pictorial means*, tended toward what we recognize as maps. Although these two traditions increasingly diverged, for numerous generations they were not readily distinguishable. The use by the Mixtec of strings of footprints to link both *places* on the groundplane and *generations* of rulers is a case in point,[41] but as David Woodward has shown, the two traditions are not firmly separated, even in the European tradition, until the dawn of the modern age.[42]

In societies in which these graphic systems ceased growing or shrunk, development of mapmaking likely slowed or ceased as well. This is in accordance with Jerome Bruner's insistence "that cognitive growth in all its manifestations occurs as much from the outside in as the inside out," and his observation that "one finds no internal push to growth without a corresponding external pull, for, given the nature of man as a species, growth is as dependent upon a link with external amplifiers of man's powers as it is upon those powers themselves."[43] By positing growth as the engine driving development and so producing history, I am insisting that the three threads twined together in *my* experience are indeed incapable of being meaningfully teased apart in human experience generally. In growing societies, the continuing need for increasing hierarchic integration produces first a simple enlargement of the mapping function, but then it's ceaseless branching. Thus the state, in its premodern and modern forms, evolves together *with* the map as an instrument of polity, to assess taxes, wage war, facilitate communications and exploit strategic resources. In Brian Harley's words, "Stability and longevity quickly became the primary task of each and every state. Against this background, it will be argued that cartography was primarily a form of political discourse concerned with the acquisition and maintenance of power."[44]

Smaller, simpler, face-to-face societies have no need to map land ownership, tax assessment districts, the topography of tank attacks, subsurface geology likely to contain oil, sewer lines, crime statistics, congressional districts or any of the rest of things we find ourselves compelled to map. This doesn't mean they don't create in their heads dense, multilayered, fact-filled maps of the worlds they live in.[45] Writing recently of the Mayoruna and Maku, the Arara and Parakana, the Arawete and the Guaja—Brazilian "peoples so remote and little known that few outside their immediate geographic area have heard of them"—Katherine Milton has observed that although life may for a while revolve around the village:

> Sooner or later every group I have worked with leaves, generally in small parties, and spends weeks or even months traveling through the forest and living on forest products. Throughout the forest there are paths that

> the Indians know and have used for generations. They travel mainly when wild fruits and nuts are most abundant and game animals are fat, but families or small groups may go on expeditions at other times of the year as well.[46]

Such peoples carry everything with them at such times, but:

> The most important possession the Indians carry with them, is knowledge. There is nothing coded in the genome of an Indian concerning how to make a living in a tropical rainforest—each individual must become a walking bank of information on the forest landscape, its plants and animals, and their habits and uses. This information must be taught anew to the members of each generation, without the benefit of books, manuals, or educational television. Indians have no stores in which to purchase the things they need for survival. Instead, each individual must learn to collect, manufacture, or produce all the things required for his or her entire lifetime.[47]

In keeping with what I would imagine of a people who did not write, Milton also notes that "tropical-forest Indians talk incessantly, a characteristic I believe reflects the importance of oral transmission of culture."[48] Others have made similar observations for groups as far-flung as the Zaire Ituri and the Aboriginals of Australia; and indeed the converse—the silence demanded in our culture by the private act of reading—has been increasingly the subject of attention.[49]

Mapmaking cultures differ from non-mapmaking cultures by the need, among others driven by mapmaking, to fill in the blanks within and without maps. In *The Heart of Darkness*, Conrad has Marlowe say:

> Now when I was just a little chap I had a passion for maps. I would look for hours at South America, or Africa, or Australia, and lose myself in all the glories of exploration. At that time there were many blank spaces on the earth, and when I saw one that looked particularly inviting on a map (but they all look that) I would put my finger on it and say, "When I grow up I will go there."[50]

Observations like this have been held up for us as examples of the power of maps to *stir the imagination* (beguiling alibi). J. K. Wright opens a widely cited paper with these words:

> *Terra Incognita:* these words stir the imagination. Through the ages men have been drawn to unknown regions by Siren voices, echoes of which ring in our ears today when on modern maps we see spaces labeled "unexplored," rivers shown by broken lines, islands marked "existence doubtful."[51]

Though he went on to qualify the songs the Sirens sang, his attention remained with the poetic quality of the song:

> The Sirens, of course, sing of different things to different folk. Some they tempt with material rewards: gold, furs, ivory, petroleum, land to settle and exploit. Some they allure with the prospect of scientific discovery. Others they call to adventure or escape. Geographers they invite more especially to map the configuration of their domain and the distribution of the various phenomena that it contains, and set the perplexing riddle of putting together the parts to form a coherent conception of the whole. But upon all alike who hear their call they lay a poetic spell.[52]

It would not just be churlish to deny this spell, it would be wrong. Human motivation is neither simple nor singular, and the purest of pleasures is often found in company with the basest of motivations; but the spell cannot be permitted to blind us to the overwhelming temptation of the *material* rewards to be reaped from the exploitation that exploration—the quest to fill in the blanks to which mapmaking cultures are driven—inevitably . . . *opens up.*

It is precisely this *opening up* (but also the *closing* such an opening implies) that attracted the attention of Brian Harley. Referring explicitly to Marlowe's remarks (and siting them in the context of Victorian colonialism) he observed:

> The passage is often quoted as an example of how maps stir the geographical imagination. But it also demonstrates the map's double function in colonialism of both opening and later closing a territory. I shall argue that Conrad's thirst for the blank spaces on the map—like that of other writers—is also a symptom of a deeply ingrained colonial mentality that was already entrenched in seventeenth-century New England. In this view the world is full of empty spaces ready for taking by Englishmen.[53]

This historical relationship was not symmetrical. The mapping—but not mapmaking—New England Indians *may* have made maps (indeed they made many for Europeans[54]), but they made none with blanks on them indicating *terrae incognitae* in the interior of the British Isles. This asymmetry manifested, *as it undergirded,* the fact that the Indians were in the English orbit as the English were not in the Indians'. It is precisely here that the alibi *of the accurate* displaces the alibi *of the Siren's song,* so that even today we are distracted from the asymmetry *of inclusiveness* by squabbles over whether it was Indian or European misinformation that finally shaped . . . this feature . . . on that map. Meanwhile, magical sleight of hand, Europeans—accurate or not—are making off with the shop.[55]

And *this* is the history of cartography! Transfixed, as professional cartographers so often are, by the minutia of projection and scaling, generalization and symbolization, it must be tempting to view the history laid out in Mesopotamian clay tablets and descriptions of Ptolemaic imagery, *portolan* charts and Renaissance engravings, thematic lithographs and computer displays of satellite transmissions, as nothing more than a halting but unstoppable progress toward an unachievable Nirvana of perfect accuracy. Certainly in such terms the maps *do* get better. There is no need to deny the staggering achievements of Eratosthenes and Ptolemy, Ortelius and Mercator, Harrison and the Cassinis. But what is masked by the bland assertion about increased "comprehensiveness of the map content" that all but invariably encrusts this historical *essence* is the way mapping—but not mapmaking—peoples get lost in the process.

Look: what of the world *as we have come to know it* is embraced on this clay tablet from Nuzi? A few acres in the Middle East? And on this Ptolemy? We have pulled back, but not shifted the center of vision. It is the same gaze, the same perspective. Marco Polo pushes the boundary east. With the Columbian encounter, more of the world is snatched up in the west. Gradually the world as we have come to know it appears, and then—inexorably, century after century (it still goes on)—what we have is the slow plugging of the holes. It is like watching a computer fill in an outline drawing with color: line by line and pretty soon . . . *none of the white is left*. Or, in the case of the world . . . *none of the red*, as Indian places become increasingly hard to find on a Ptolemaic grid littered with . . . New Londons and New Spains:

> For seventeenth-century New England, the map is a text for studying the territorial processes by which the Indians were progressively edged off the land. I am not suggesting that maps were the prime movers in the events of territorial appropriation and ethnic alienation. My contention, however, is that as a classic form of power knowledge maps occupy a crucial place—in both a psychological and practical sense—among the colonial discourses which had such tragic consequences for the Native Americans.[56]

And now *everyone* is on this map, *everyone* has been caught up in this panoptic gaze.[57] As time passes more and more of the world gets caught up in a view from a center that shifts only slightly, from Mesopotamia to Greece and Rome—around the Mediterranean—up to France and England (momentous shift of a couple of hundred miles). *The prime meridian?* Of course it runs through Greenwich! But it has to run . . . *somewhere*.

Alibi of the necessary: it doesn't have to be . . . *there*.[58] It must be insisted on that this largely is not a disinterested cartographic activity, but the result of the same intertwining of polity and mapmaking we have referred to before, an activity required for the stability and longevity of the state. This is to say that, in a very important sense, the map *requires and justifies* as it *records and demonstrates* transformations in control over the land, its appropriation in the name of science and civilization, the state and human progress (indeed it is precisely this process of appropriation and the consequent acculturation that is the focus of Milton's research among the Mayoruna and the Maku). Mapmaking societies . . . *reach out*, not of course to make maps more comprehensive (much less more truthful), but in the unfolding of the dynamic that their growth and development have helped to set in motion (and in which the cartographic enterprise is an essential and committed partner). In so doing they subsume whatever they can—the labor and other culture of those they encounter—and in this way their growth is fueled and their development pushed from without (that is, by conquest, appropriation and seduction) as well as from within. Stripped from those . . . *ripped off* . . . is not only their place, their energies, their knowledge about plants and animals, but their language, myths, rituals, customs and artifacts. It is not only explorers, missionaries, soldiers, slavers, trappers, miners, loggers and colonists who have encroached on such peoples, but anthropologists and their predecessors.[59] Characteristically, I am able to pull from my shelves, *Xingu: The Indians, Their Myths* (as the Xingu are unable to pull from their shelves anything about me) and read from its dustjacket that "as a source of ethnographic data for structural and comparative analysis, [the myths] are invaluable."[60] There is nothing we of the ever-growing mapmaking societies will not take and make use of. This way these great developing cultures—the "West," the "East," the "Islamic nation"—increasingly differentiate themselves from the less developed societies and cultures they ever more voraciously consume. What distinguishes "the West" most tellingly from the Kamaiura or the Ainu or the Navajo is not this view or that, but that in "the West" . . . *there are so many views*, that whatever it is—the origin of the world, or the relationship between man and nature—*is seen so many more ways* than it is among the Hopi or the Bororo or the Inuit. In the end there are not just so many more maps, but so unfathomably many more *kinds* of maps. It is in this way that mapmaking fuses its growth and development with history, in the transformation of the world from a mosaic of peoples to a mosaicked people.[61]

A way in which I, driven by my own growth and development (interactively with the history of those around me), fathered two sons, who even as I write are sitting in the next room glued to a monitor where, in SimCity, thematic maps flash onto the screen to record and embody the "city" my sons are attempting to create. I am not terribly happy about this. I would rather they were out in the woods, or if not the woods then exploring the city whose pavement runs hard beneath their feet. But in the mapmaking society we live in this is what it's come to.

CHAPTER THREE

Every Map Shows This . . . But Not That

Are there no consequences of this? Is nothing implied by the mouse-summoned maps flashing on the computer screen? As the kids play, they call first for this image, then that. Pollution, population growth, fire coverage, crime—Chandler brings up each in turn. Randall fine tunes the traffic patterns—*flash!* traffic grid *flash!* traffic density. "Look how big my city is," says their good friend Kelly, as he pulls back . . . for a regional perspective.

Power grid, population density, police coverage, land values—what's missing? what's been omitted? Are not these precisely the riches members of mapmaking societies have piled up? Is not this the point, this evidently beneficial access to every kind of map? Here, in SimCity (a game for *kids*): a baker's dozen dynamic, interactive maps! Are not these the benefits accrued from standing on the shoulders of giants, giants who themselves stood on the shoulders of giants . . . or on the shoulders of even quite ordinary men and women—for given the generations of accumulated culture laid up and garnered here, what difference could a couple of feet amount to? It is this *subsumed and amassed cultural capital* that mapmaking societies bring to the task of making maps; not the patiently acquired mastery of this or that individual more or less carefully passed on—often in secret—through speech or gesture or inculcated habit. It is the endlessly reproduced and everywhere disseminated wisdom of thousands of such individuals, caught up, stored, annotated, corrected, indexed, epitomized, reduced to formulae, taught by rote, so that what once was an epochal discovery or invention is reduced to common knowledge, ground into a taken-for-granted fact of life. What enables high school students to say, "169," within a millisecond of hearing, "What's 13 times 13?" is not some superior intelligence vis-à-vis their pre-Socratic forebears, or their hunting and gathering contemporaries in the rainforests of Brazil, or even their former classmates who dropped out in the seventh grade, but precisely *access* to this hoarded wealth of cultural labor heaped up . . . *and mined.*

Surely then the claims we might make—as beneficiaries of this enormous trove of learning and knowledge and technology—to accuracy, to comprehensiveness—*to the truth, damnit!*—must be greater than those our predecessors could make, shackled as they were by . . . by what? Well, by their lack of satellites, for instance, satellites such as those exploited by Tom Van Sant to make this image—famous now, famous *already*, this . . . what to call it? *portrait map*—of the earth.

Portrait map: that's what everybody calls it. Here, in its first publication in the November 1990 *National Geographic*, it is headlined, "First-of-a-kind portrait from space."[1] In the new *National Geographic Atlas of the World*, which the article was promoting and where the image appears on the half-title page, the title page and the copyright page, it is captioned: "The global portrait . . . " and "this portrait."[2] On a widely distributed poster it is simply called . . . "A *Clear Day*."[3] This obliqueness is appropriate for a map that would like us not to notice the fact that it is a map, that would like us to imagine that this is what we would *see* were we to *look*—for example, from a satellite—down onto the earth, that would like us to think of it . . . *as a portrait*, that is, as a picture, painted, drawn or photographed from . . . *life*.

What is this then but the acme of cartographic perfection, an image of the world so . . . *true* as to render all the questions raised in the preceding chapters about perspective, authorship, point of view—and all the rest of the revisionist claptrap—hopelessly *out of date?* Here it is—*finally*, after all these centuries—the next best thing to an eyeball in space! Of course it's not. This is not what L. Gordon Cooper saw on that final Mercury flight in 1963 when mission control thought he was hallucinating because he told them he was seeing things they assumed he couldn't, things like trucks on dirt roads and a train on its tracks and buildings with smoke pouring from them.[4] None of these things is visible in this portrait—even under a lens—for what we have here is nothing that can be seen by an eye, but precisely that reality which we know a map alone can give us—that reality which *exceeds* our vision, our reach, the span of our days—but which at the same time, we also know, is therefore *necessarily* a product of social construction. Yet so accustomed are we to seeing the earth portrayed this way—the bright land against the dark blue oceans, its symmetrical disposition around a central Atlantic, Antarctica a band of white across the bottom—that nothing seems problematic about this image. It seems to be the same earth, just . . . *more real, more accurate, more . . . true to life.*

Like the photograph the poster claims it to be. Yet how seriously can we take this claim? Roland Barthes asks:

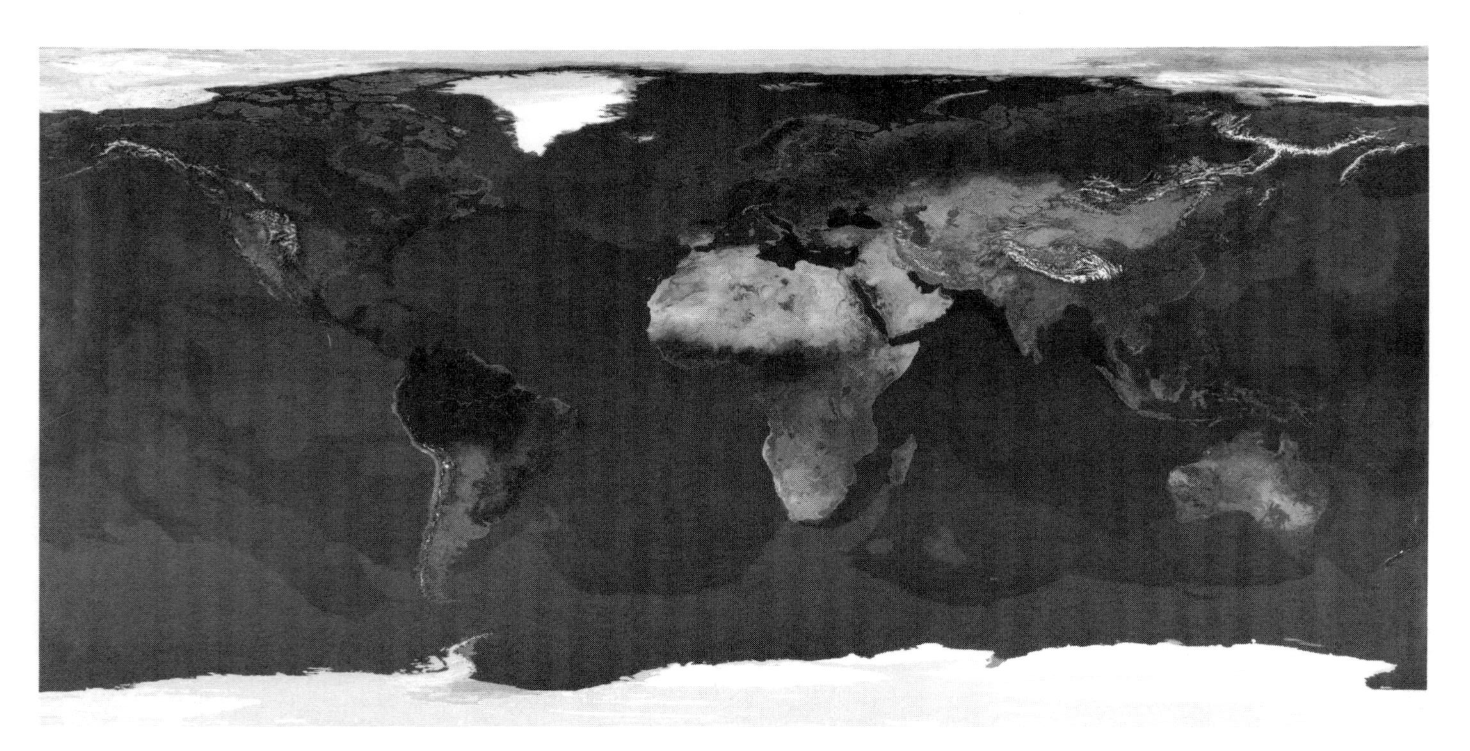

Satellite composite view of earth. (Photo credit: Tom Van Sant/The GeoSphere Project, Santa Monica, CA)

> What is the content of the photographic message? What does the photograph transmit? By definition, the scene itself, the literal reality. From the object to its image there is of course a reduction—in proportion, perspective, color—but at no time is this reduction a *transformation* (in the mathematical sense of the term). In order to move from the reality to its photograph it is in no way necessary to divide up this reality into units and to constitute these units as signs, substantially different from the object they communicate; there is no necessity to set up a relay, that is to say a code, between the object and its image. Certainly the image is not the reality but at least it is its perfect *analogon* and it is exactly this analogical perfection which, to common sense, defines the photograph. Thus can be seen the special status of the photographic image: *it is a message without a code . . .*[5]

This is a generous description of the content of the photographic message: "literal reality," "perfect analogon," "message without a code." Barthes asks only that the object not be cut up; that there be no code intervening between the object and its image; and that the image not constitute a mathematical transformation of its object. Can the Van Sant meet these conditions?

The Dividing Up of the Reality

None of them. In the first place it was necessary to divide up the reality into units—hey! the earth is enormous—into thousands of units in fact, even if we stay only on the level of the "thousands of satellite photographs" the poster caption on *A Clear Day* refers to. But in the words of the *National Geographic Atlas:*

> This portrait was created by artist Tom Van Sant and scientist Lloyd Van Warren of NASA's Jet Propulsion Laboratory from visible and infrared data recorded between 1986 and 1989 by National Oceanic and Atmospheric Administration satellites. Orbiting at an altitude of 850 kilometers, the satellites scanned the surface in four-square-kilometer sections, or pixels.[6]

How many pixels? *Thirty-five million* of them.[7] This is less a *dividing up* than a grinding or pulverizing of reality, done in this case with multispectral scanners. Scanners don't take photographs or snapshots, but *scan* the land by virtue of a mirror "that sweeps back and forth in a motion likened to a person sweeping with a broom while walking across a room."[8] The mirror directs the light reflected from the earth through a prism onto a small array of photoelectric or other detectors. The modulated electric signal these generate—millions of bits of data per second (no mortar and pestle grinds so fine)—is then transmitted to earth in digital form where it is recorded and later computer-processed to produce photographs.[9] Though elaborate, this process sounds straightforward. It's not. Each step not only determines a perspective through which to view reality, but resulted from a choice that was (usually) . . . bitterly contested. That is, the first intrusion into Van Sant's map of the society out of which it emerged occurs at the level of . . . *the scanner.*

Van Sant's map was constructed from imagery collected by National Oceanic and Atmospheric Administration TIROS-N series satellites. The "T" in TIROS stands for "Television" (TIROS is an acronym for Television and Infrared Observation Satellite), an historic residue of the early days when TIROS used television cameras—or return beam vidicons—to transmit what were then *astounding* images of clouds.[10] But television didn't last. In his brief history of the Landsat program—for which the multispectral scanner was developed and in which it was first used—Stephen Hall emphasizes the battle waged by proponents of the scanner against those advocating expanded use of return beam vidicons. Hall's heroine is Virginia Norwood, the scanner's principal designer (and indeed RCA's Landsat vidicons did fail within hours after launching). But in picking up the battle where he does, he ignores a third option, also closed to TIROS, the return-film satellite. Developed by the CIA and first successfully deployed in 1960, 86 such satellites had been orbited by 1976: "These satellites took photographs of the earth during a few days of orbits and then ejected the film in a special capsule. The capsule reentered the atmosphere, descending by a parachute, and was recovered in midair by an airplane towing a device that snagged the lines of the parachute."[11] The advantages of return-film satellites were straightforward: they "could provide finer-resolution images with less distortion than either a vidicon or scanner and would, therefore, be particularly useful for mapping."[12] Why not use them? Aside from the fact that "the Bureau of the Budget refused to allow NASA even to list mapping as an objective of Landsat because the Department of Defense already utilized classified satellites for mapping" and "for NASA to do the same thing would be a duplication of effort," the Department of Defense finally simply refused to allow NASA to do so. This is to say that the available window of fine resolution and low distortion was closed by . . . *military paranoia.*[13]

The Landsat scientists who in the end "decided that [a film-return satellite] would not be in the best interest of the country as a civilian experiment"[14] hoped to compensate for the deficiencies posed by the scanners and return beam vidicons with repeat coverage and multispectral data. Since TIROS satellites ended up carrying multispectral scanners, the Landsat experience is relevant here too. In this earlier case, which windows would be opened was contested by NASA, the

Department of the Interior (especially the Geological Survey) and the Department of Agriculture:

> Different sets of spectral bands were useful for different disciplines, so NASA had to negotiate a compromise among scientists in different disciplines both within NASA and the user agencies. This disagreement came to a head in a May 1969 meeting of NASA scientists working with scanners at which "there was general agreement on the types of scanners needed but serious disagreement on the wavelength intervals to be used." The spectral band selected by the Goddard study for the scanner on Landsat followed the recommendations of the Department of Agriculture and had been agreed to by the Geological Survey, which had little interest in that sensor. [They wanted film if possible but if not, then return beam vidicon data.] The scientists working with aircraft scanners at the Johnson Space Center disagreed, claiming that not enough was known to select satisfactory spectral bands for a scanner on a satellite. In that case the primary issue revolved around NASA scientists, who wanted to do more research.[15]

The resulting compromises satisfied no one, and Pamela Mack concludes that their real effect was to strengthen NASA control of the Landsat program by weakening the argument for any one user agency to take over its operation.[16] This is critical because although NOAA would end up operating the TIROS system, the initial TIROS-N was developed and lofted by NASA. As in *every* case, the spectral windows opened on the world were selected for . . . *bureaucratic advantage*.

More of these windows were opened on TIROS than on the early Landsats, but each remained glazed . . . *by the data processing system*. The sensor payload on the N series included the Advanced Very High Resolution Radiometer, a High Resolution Infrared Sounder (despite the names, resolution remained constrained by military considerations), a Stratospheric Sounder Unit, a Microwave Sounder Unit, a Data Collection System and a Space Environment Monitor.[17] Interagency squabbling—inevitably—contributed to the Rube Goldberg contraption NASA created to deal with the unprecedented volume of data and the disparate demands of its users; but ultimately, it was "*political constraints, particularly those set by the Bureau of the Budget [which] shaped the choice of the data processing system.*"[18] Anyone in the world equipped with appropriate equipment can receive real time output from the Advanced Very High Resolution Radiometer, but for central processing, data are received only at stations in Alaska and Virginia (which means that it is recorded and then played back *on the satellite*), whence they are retransmitted to a specialized Digital Data Handling facility in Suitland, Maryland. Here they are processed first on dedicated medium-scale computers, then by hand or on very large computers, before being recorded on magnetic tape, an electron beam recorder, or both. Any notion that these products have not been *extensively manipulated* even before Van Sant begins to fool around with them can therefore be dismissed out of hand. Here: this is some of what has happened to the data even before it has left the satellite:

> The digital outputs of two selected AVHRR channels are processed in the manipulated information rate processor (MIRP) to reduce the ground resolution (from 1.1 km to 4 km) and produce a linearized scan so that the resolution across the scan is essentially uniform. After digital processing, the data are time multiplexed along with appropriate calibration and telemetry data to an analog signal, low-pass filters the output, and modulates a 2400-Hz subcarrier. The maximum subcarrier modulation is defined as the amplitude of gray scale wedge number eight, producing a modulation index of 87 ± 5 percent.

TIROS benefited from experience gained during the development of Landsat (where the data processing system was not only slow and unreliable, but produced images so degraded few users wanted them) and thereby generated improved products, but even so, the data-processing window remained fogged by . . . *political constraints*.

The Code Between the Object and Its Image

Smashed to smithereens, shamelessly manipulated: evidently endless codes slither between the object and its image. The light reflected from the earth is broken up, it is transformed into an electric signal, it is recorded, it is played back, it is transmitted, it is received, it is recorded again, it is calibrated, annotated and recorded yet a third time, or it is calibrated, annotated and used to drive an electron beam recorder to produce . . . a picture. Each of these steps reflects a code that says, *this* means *that*. Take the prism in the scanner. Essentially this *filters* light over detectors "tuned" to different wavelengths. In the original Norwood model there were six of these; in the early Landsats only four; in the Thematic Mapper, seven; in the TIROS-N/NOAA satellites, four or five.[19] These wavelengths were not chosen at random but because alone or in combination they corresponded to the existence of particular things on the earth:

> Early work by the Department of Agriculture, using aerial photography, had shown scientists that differentiating crops would not be easy. At many stages of growth, small grains such as wheat, barley, and oats are almost identical in color. However, if the color of the plant is broken into a spectrum, showing how much of each color of light the plant reflects, a characteristic pattern called a *spectral signature* emerges.[20]

Before these signatures could be interpreted, a lot of *ground truthing* had to take place. What this means is that someone had to take a photo into the field (that is, *down onto the ground*) and correlate the image with what was found there. Once it was deciphered it proved to be a code about as straightforward as that of the traffic system, one that, in its simplest version, is increasingly familiar to anyone who consults an atlas. Here's the code as laid out in a caption in the *National Geographic Atlas of North America:* "Vegetation appears as red, built-up areas as bright blue, clouds as white, water and cloud shadows as blue–black."[21]

In such a "false color" image, the coding could not be clearer: red = vegetation, blue = built-up areas, white = clouds, and blue–black = water and cloud shadows.[22] Originally developed during World War II (enormous surprise), false color photography was used to distinguish *live* vegetation—which appears red on false color film—from *dead* vegetation (or paint) used as camouflage—which appears blue or purple. The difference results from the different emulsions used, false color film being sensitive to red, green and infrared wavelengths of light instead of the more usual red, green and blue. With digital data, however, the assignment scheme is not dependent on the emulsion. Any code, however arbitrary, is possible. In our case: "Van Sant chose colors that would give a realistic view of the physical world. Gray–brown areas along the coasts represent silt discharges of great rivers, algae blooms, or upwellings of cold, deep water."[23]

A *realistic view?* What can this mean but that since the film cannot be relied on, *humans will apply the appropriate colors?* But then, what was the point of using the photographs in the first place? Already manipulated to remove "distortions," these *hand tinted* images increasingly resemble the paintings they seemingly took such pains not to be. The entire objectifying apparatus—35 million pixels, computers, satellites, precision-processing for geometric accuracy—comes increasingly to resemble nothing so much as a painter's kit—easel, palette, brushes, knives—in which context "realistic" takes on a different sound. "In the vaguest and most general sense the term is often applied to works which instead of choosing conventionally beautiful subjects depict ugly things or at least scenes from the life of the poorer classes," is the first meaning given in *The Oxford Companion to Art.*[24] Other definitions contrast "realistic" to "abstract," "distorted," "stylized," "idealized" and "generalized," The first thing we read in *The Encyclopedia of Philosophy* is that in medieval thought the "term 'realism' was used, in opposition to nominalism, for the doctrine that *universals* have a real, objective existence. In modern philosophy, however, it is used for the view that material objects exist externally to us and independently of our sense experience," that is, in opposition to idealism and phenomenalism.[25] The *Princeton Encyclopedia of Poetry and Poetics* does talk about giving "a truthful impression of actuality as it appears to the normal human consciousness," but the discussion here is as crabbed as elsewhere.[26] The fact is that "realistic" is a loaded word, a word which several thousand years ago lost any innocence it might once have claimed. A *realistic view* . . . evidently the colors are not only coded but thereby plunged into every debate concerned with being and existence.

The Mathematical Transformation of the Object

Chopped up, endlessly coded: on its way to the earth, the image is also subjected to a transformation in precisely the mathematical sense of the term. The trip from sphere to plane requires this, it is not something done for fun, at least not by cartographers, who almost to a person lament the inevitable deformations and contortions, distortions and misrepresentations that are the unavoidable consequence of attempting to display simultaneously on a plane the entire surface of a sphere. Waldo Tobler has said, "A map projection can be considered a transformation applied to spatial point coordinates,"[27] that is, as the substitution of one configuration (as by rotation or translation) for another in accord with a mathematical rule, that is, in accord with a code which says, *given this, then that*. In *Theoretical Geography* Bill Bunge put it this way:

> Imagine the following physical equipment: a blackboard on which is drawn a representation of the earth's surface, a portable bulletin board with the opaque cork board replaced by a plate of glass and this contraption placed about twenty feet in front of and parallel to the blackboard. In addition, imagine a large number of strings, each string having one end glued to the plate of glass so that it is not difficult to imagine that each point on the blackboard map has a string connecting it to each point on the glass. The strings establish a *one-to-one correspondence* between the blackboard and the glass. The particular relationships of the set of points at the blackboard end of the strings to the set of points at the at the glass end of the strings determines the *transformation* or the geometric rules under which we are constrained to move from one surface to another.[28]

Only now imagine that instead of a representation of the earth on a blackboard it is a spherical globe from which the strings must be projected. *What a mess!* It's easy to imagine the strings from the front of the sphere falling on the glass, but how about those from the back. Should this string from way in the back come around from the left . . . or the right? Should it come up over the globe . . . or from underneath? The questions change as we change the surface to which we run the strings. The plate of glass doesn't have to start off flat (when it is, the projections are planar): the glass could be wrapped around the globe and later slit open and rolled flat (this would result in a cylindrical projection). We could do something similar with a cone or with this or that polyhedron (resulting in conic and polyhedral projections) or, in fact, with any developable surface or set of surfaces (polyconics and polycylindricals and so on).

But there's no stopping this, we don't have to be able to visualize the surface, the transformations can obey any rule, they can be endlessly combined, the possibilities stretch on forever. "An infinite number of distinct projections are possible," writes Tobler.[29] Robinson *et al.* say, "An infinite number of map projections is possible."[30] "There are an infinite number of possible map projections," intone the editors of *Goode's World Atlas*.[31] How to chose? This is *the* question, for the answer determines the way the earth will look on the map. How different can this be? Well, in a Lambert Azimuthal Equal-area projection centered on the north pole, the pole is a point. Of course the further you get from this point, the more weird everything looks (at the edge of such a map Australia and Antarctica are so long and skinny as to be hard to recognize). On the other hand, on a Mercator projection centered on the equator, the pole can't be shown at all; it's turned into a line of *infinite* length, so the closer you get to the pole the greater the areal distortion (thus though actually one-fifth the size of Brazil, Alaska appears on the Mercator to be the same size). Yet no projection is without its advantages. The Mercator may distort areal relationships, but it preserves shapes (it's conformal), and it's the only projection on which loxodrones (lines of constant compass bearing) are straight. Therefore it's widely used for charts.[32] Equal-area projections, on the other hand, are essential for displaying things like population, vegetation, crops, religions, and other distributions.

And *every* projection is like this, good for one thing, but not another. As Wellman Chamberlin puts it, "One must choose between equal-area scale and conformality. These two most important qualities in map projections are mutually exclusive. The same map cannot have both," because he goes on, "equivalence of area is maintained by decreasing the scale in one direction as it increases in another. In conformal maps the scale changes equally in all directions so that any small portion of the map has its correct shape."[33] Each quality is valuable, but for different things. What this means is that the selection of a map projection is always *to choose among competing interests*; that is, to *embody* those interests in the map . . . even if we confine ourselves to such superficially technical issues as the representation of angles and areas, distances and directions.

It is easy to pretend this isn't so, to act as if the choice were an "objective" one, that somehow it were possible to . . . *rise above interest*.

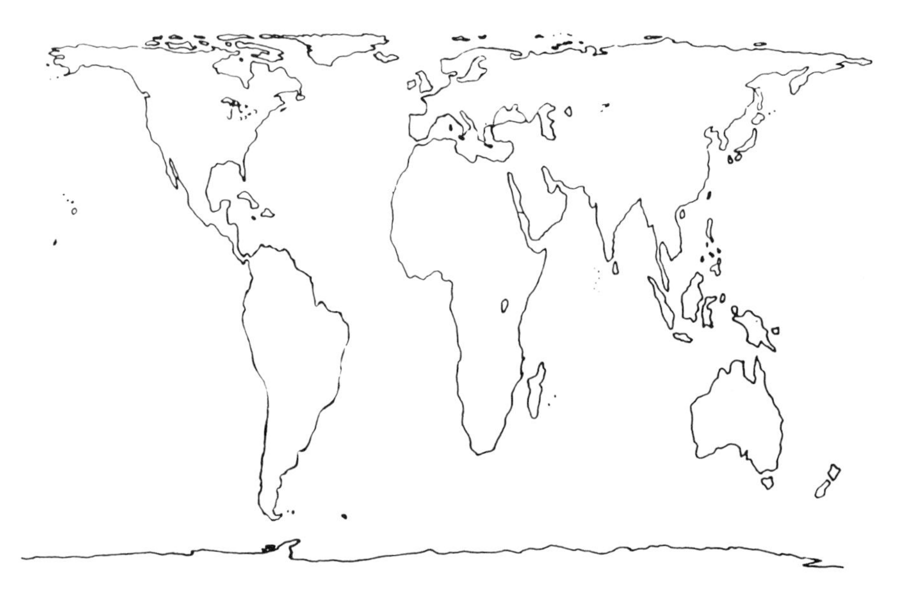

The equal-area Peters projection. (*World Map: Peters Projection* by Arno Peters. Copyright Akademische Verlagsanstalt. Distributed in the United States by Friendship Press, New York. Used by permission.)

In explaining the decision of the *National Geographic* to adopt the Robinson projection for its world maps, Chief Cartographer John Garver, concludes, "The projection does not espouse any special point of view, and we believe that its compromises are the most reasonable for a general reference map of the world."[34] *Sans doute*. This would be the easier to accept if John Garver had not made such a big deal about the personal reasons for his satisfaction with this decision (former graduate student of Robinson's *et cetera et cetera*), and if Robinson had not been the leading voice in a perfidious and vitriolic attack on a competing projection.[35] What is at stake? Certainly nothing . . . "scientific."

The object of Robinson's attack? The Peters projection. In the index to Mark Monmonier's tendentious little *How To Lie With Maps* this is cross-referenced to . . . the *Gall*-Peters projection . . . where we read (*how to lie with indexes*): "not first equal-area projection, 97; preservation of area and distortion of shape by, 97–98; used in media campaign, 98–99."[36] These entries epitomize the concerns of the American academic cartographic establishment, that German historian Arno Peters made an unjustified priority claim,[37] that his projection is ugly,[38] and that it was adopted by the World Council of Churches, the Lutheran Church of American and numerous United Nations and other international agencies only because "Dr. Peters knew how to work the crowd."[39] What Peters was on about was the fact that the most popular projections consistently exaggerated the size of the higher latitudes—in effect the land masses of the northern hemisphere since that's where most of the land is—at the expense, not only of some version of the truth (double-bladed sword), but of the self-image of the developing world. Hall is good here:

> By correcting one distortion, projections inevitably create another, and the historical evolution of map projections is like a mathematical shell game that always seems to cheat the southern hemisphere. In the Van der Grinten projection, used by the National Geographic Society between 1922 and 1988, some parts of the globe were wildly out of scale; Greenland, long the bane of cartographers because its high latitude incurs spatial exaggeration, was 554 percent larger than actual size, the United States 68 percent larger. On the Robinson projection, which improves considerably upon Van der Grinten, the Greenland exaggeration is only 60 percent. Less wrong, but wrong nonetheless. Africa, significantly, is 15 percent *smaller* on the Robinson projection. As cartographic expert John Synder puts it, the Robinson projection was selected because it offered "the best combination of distortions."
>
> Why does it matter? Such errors can have an impact out of all proportion to their size, as the American Cartographic Association well understands. "A poorly chosen map projection can actually be harmful," the association recently noted. "We tend to believe what we see, and when fundamental geographic relationships, such as shapes, sizes, directions and so on, are badly distorted, we are inclined to accept them as fact if we see them that way on maps."[40]

Peters did more than insist that whatever its appearance *an equal-area projection* was the only fair way to show most things worth showing about the world. He implied—no, he *pointed out*—that the use of most other projections had a powerful built-in bias. Here's Hall again:

> Peters argues that the Mercator projection has promoted the "Europeanization of the earth," and that the customary practice in atlases of using many different scales to show different parts of the world is literally belittling to Third World nations. Terry Hardaker, chief cartographer of the *Peters Atlas*, goes further. He has written that other map projections offer "the equivalent of peering at Europe and North America through a magnifying glass and then surveying the rest of the world through the wrong end of the telescope."[41]

David Turnbull asks questions to a similar end: "If you compare the Mercator projection with the Peters projection, a map which endeavours to preserve relative size, what differences do you discover which might have cultural or political significance?" He explicitly pushes us to ask what *interests* might be served by the use of a Mercator projection: "Is it a coincidence that a map which preserves compass direction (a boon for navigation) shows Britain and Europe (the major sea-going and colonizing powers of the past 400 years) as relatively large with respect to most of the colonized nations?"[42] Not only does the cartographic establishment take umbrage at the implication of their complicity in these nefarious imperialist activities (they labor only for science), but they object to the emphasis on the Mercator projection, which they point out is 400 years old and has been succeeded by, as we know, an all but infinite collection of alternatives. Therefore, they insist, the "Peter's approach is more propaganda than science."[43] But as we have already seen, the attention to "propaganda" is an alibi. It does nothing but deflect attention from the fact that the selection of *any* map projection is *always* to choose among competing interests, is *inescapably* to take—that is, to *promote*, to *embody in the map*—a point of view. Robinson's is essentially . . . *aesthetic*. His description of the Peters projection is that of an art critic—"wet, ragged, long winter underwear hung out to dry"—not a scientist. This is fine. Robinson has always arrogated to himself the mantle of the artist.[44] But if it is only with respect to the *aesthetic* dimension of the continents that his projection bests the Peters, it is not only difficult to justify the *Geographic's* ennoblement of the Robinson, but to understand what Garver is talking about when he characterizes it as matching "reality more closely than its venerable predecessor."[45] *In what way?* No one claims the Peters departs from reality—*it just looks funny*.[46]

It really is a shell game. When the aesthetic issue gets hot, switch to science and talk about accuracy, but when that bluff is called, bring on the "wet, ragged, long underwear." But as Brian Harley has testified, it's a shell game that is played for keeps:

> Yet cartographers, though they are fully aware how maps *must* distort reality, often engage in double-speak when defending their subject. We are told about the "paradox" in which "an accurate map," to "present a useful and truthful picture," must "tell white lies." Even leaving aside the element of special pleading in this statement (the map can be "truthful" and "accurate" even when it is lying), there is the corollary that cartographers instinctively attribute the worst forms of "ignorance," "blunders," and "distortions," and so on to non-cartographers. For instance, when they come to talk about propaganda maps or the cartographic distortions presented by the popular media, a quite different order of moral debate is entered into. The *cause célèbre* of the Peters projection led to an outburst of polemical righteousness in defense of "professional standards." But ethics demand honesty. The real issue in the Peters case is power: there is no doubt that Peters' agenda was the empowerment of those nations of the world he felt had suffered an historic cartographic discrimination. But equally, for the cartographers,

> it was their power and "truth claims" that were at stake. We can see them, in a phenomenon well-known to sociologists of science, scrambling to close ranks to defend their established way of representing the world. They are still closing ranks. I was invited to publish a version of this paper in the ACSM *Bulletin*. After submission, I was informed by the editor that my remarks about the Peters projection were at variance with an official ACSM pronouncement on the subject and that it had been decided not to publish my essay![47]

Van Sant's image, scanned and filtered, corrected for distortions and hand tinted, is now subjected to a mathematical transformation: it is stretched and squeezed—that is, *re*distorted—until it conforms to the aesthetic dictates of the . . . Robinson projection.[48] What can remain of the claims made by virtue of the portrait's photographic transparency to . . . *reality*, to . . . *accuracy*, to . . . *truth to life?* Savaged by military paranoia, "realistic" color-coding, bureaucratic infighting, aesthetically motivated remanipulation, politically motivated budget cutting . . . it cannot be much.

"Night and Day, You Are the One"

But let's say there had been no problems with the choice of spectral bands, that the Department of Defense had permitted the use of high-resolution images with terrific geometric accuracy, that the data processing had been ideal, that the colors had not been recoded, that the projection chosen had been . . . the Peters (there is no getting around this, some projection must be used, any involves a choice among competing interests) . . . that the image did *indeed* conform—at the level of the photograph—to Barthes' . . . message without a code . . . that what we were confronting was . . . *a snapshot*.

Even a snapshot would have to be taken . . . *at some time*, and whenever that was, half the earth would be in darkness, or nearly half, fringed "where the blue of the night meets the gold of the day." Bing Crosby wouldn't have recognized this daytime-only world which we have bought into for pages. Why should he? The map displays a self-evident impossibility: for all the plausibility of its "natural-looking colors" . . . the earth as it might appear could it be illuminated *all at once*, a construct, an artifact, an invention . . . *on its face*.

Here, then, a counter-image, the Hansen planetarium's map of the earth at night. With bands of lacquer streaking the oceans, an artist's impression of the aurora, type right over the South Pacific, and heavy annotation, this is a map that is less interested in being mistaken for a photo. Instead of self-congratulatory puffery—"spectacular global image," "this dazzling image"[49]—a block of type headed . . . "CAUTIONARY NOTES." It's worth having in full:

> There are several aspects of this image that the discriminating viewer should keep in mind. First, it is based on a mosaic of about forty individual photographs, each of which has its own distortions. These photos have been approximately reduced to a common scale, but the final image corresponds to a Mercator projection only to about 5% accuracy. The individual photos were also taken with a variety of exposures and under varying moonlight. Several additional processing steps then led to this poster, with the result that quantitative light intensities in different regions can only roughly be judged. The photos in the mosaic were taken at various times and seasons over the period 1974–1984; this affects in particular the occurrence of tropical fires, which are highly seasonal. The aurora included in the image is an artist's rendering based on satellite photography. Its positioning is only meant to be suggestive—aurorae in fact usually occur in a ring centered on the north or south magnetic pole. Finally, although most dark regions are truly lacking in light, sections of some photos were clouded out, and suitable photos were not available for a few regions such as many remote islands, and portions of South China and southwestern Africa.[50]

Earth at Night is frankly a map, a complicated human artifact, shaped and limited by its ambitions.

What sort of a critique does *Earth at Night* present to *A Clear Day?* Without vacating the photographic terrain staked out by the Van Sant—both are smoothed mosaics of satellite imagery—it calls into question its . . . *comprehensiveness.* At the most fundamental level, the Hansen queries the implicit assumption that the Van Sant shows us the earth the way it . . . *is.* One of the reasons the Van Sant is able to suggest this so powerfully is because it seems *to conclude a history,* that of cartography, at least that version of it written as the quest for ever more "realistic" portrayals of the earth. The Van Sant is able to conclude this history by showing us the earth exactly as we've always seen it—only better. Confronted with its shimmering fragility, it is hard for one to imagine that a better world map could ever be made. Then you look at the Hansen. *What is this?* At the level of sensation, it is an image of . . . *radiated* as opposed to . . . *reflected* light. But this has the immediate consequence of forcing the conclusion that there could be . . . *many other views* . . . and that pulls the rug from any notion that the Van Sant—*and its long line of predecessors* —has any particular claim to constituting . . . *the* view.

But this difference between reflected and radiated light is not an innocent one. Reflected light comes from the sun and *bounces* back into the scanner. It has a kind of *natural* quality. But radiated light is another story, something had to burn, someone had to burn it, it is thoroughly impregnated with . . . *culture*. So what looked like a sensory difference that queried the Van Sant at the level of its data turns into a thematic difference that queries the Van Sant at the level of its information: the earth . . . *without people?*[51] Immediately one wonders—*what else is missing?*

It is a small step to the hidden. What does the day hide? *It hides the night*. And this amounts, in the end, to a denial that the earth turns on its axis. Is this an obligatory burden of a map? Not at all. But it is . . . of a snapshot.[52]

"Blue Skies, Shining On Me"

A snapshot would also have caught the enveloping coat of cloud, would have suggested at any rate the existence of that aspect of the earth that above all else differentiates it from so many of its sisters in the solar system, *its atmosphere*, rich in water vapor, condensing into clouds, moving off the oceans onto the land bringing . . . *life-giving rain*. And here, much closer to a snapshot is an image from this month's *National Geographic*, a blue–black roundel, splattered and spotted with white, hard to make out anything *but* the clouds (and a high light off the Pacific), a bit of North America peeping through, but South America . . . *obliterated*. And this *is* the earth, the one I know anyhow. The caption says, "Four images taken at 60-minute intervals on July, 11, 1991, by the GOES-7 weather satellite were combined with a global view from the satellite on the same day."[53] Compared to the Van Sant or even the Hansen—which also eschews cloud cover—this is a snapshot. Of course, what sticks out, what strikes, what cannot be missed . . . *is the atmosphere*. Without its atmosphere the earth would not be the earth, it would be lifeless, it would be another planet altogether. You might object that showing the clouds would obscure the land and sea—and of course they would—but without them the Van Sant is not a portrait of the *earth* but at best a picture of its *land* and *sea surfaces*. Painted this way, your portrait would show no more than bones . . . and blood.

"All Summer Long"

So far is the Van Sant from being a snapshot that "Data from different times of year were acquired to ensure the best lighting and maximum vegetation."[54] What can this mean but that as the map gives us a planet without night, it also gives us one without seasons? Here, I am looking at another map now, this one of average dates of last spring frost. Next to it is another of first fall frost. And here on the opposite page is a third, entitled "Climate Zones," of average minimum winter temperature.[55] The zones sweep up the map as the green sweeps up the land in the spring, or down the map as the brown drops in the fall. It is a world I recognize, one of annual as of diurnal change. What does the Van Zant come to but a map of rainfall? What does its lack of dynamism amount to but a denial that the earth is tilted on its axis, as its simultaneous portrayal of summer in the north *and* the south amounts to a denial that the earth is a sphere? Is this bad? Not necessarily. What is objectionable is the way the map wants to have it both ways, to be at once a snapshot, unmediated except for the necessary intrusion of chemical processing, and *therefore* objective, true; and *at the same time* to revel in the enormous labor of its production, the arduous sifting of mounds of images collected across . . . 3 years.

Though contemporary maps tend—or pretend—to show the distribution of things in space at—*flash!*—this instant (that is, to constitute themselves a form of snapshot), not all do (historical maps notoriously don't[56]), and the idea that maps should is no more than a prejudice. What bothers me is the way it is blandly asserted that, "Artist Tom Van Sant and scientist Van Warren chose scenes from different times of the year to ensure the best lighting and maximum vegetation,"[57] as though these were desiderata so taken for granted that they could constitute some sort of Cartographic Seal of Approval. What does it *mean* that the photos composited here were collected across a three-year period:? Why not thirty? thirty thousand? thirty million years? Through each of these windows different features of the earth would appear dynamic, others static. Through this one the continents would be seen to move, through that the great ice sheets, through the third . . .

The choice of temporal resolution is not as it might seem a matter of convenience, but a way of defining the subject. There is no reason why your portrait shouldn't highlight the fine hair you had as a baby, the exciting physique you had at twenty-five, the maturity of your features at forty, the serenity of your expression at seventy . . . but we call this a collage. Is it more . . . *true to life?* In some ways, but not in others.[58]

What Is the Map For?

Barthes not withstanding, not even a snapshot can speak for itself. It too, however minimally, is coded, needs to be . . . deciphered, needs its text.[59] *What am I looking at?* And *What am I looking at here?* And a finger descends to isolate this or that feature, and a voice in my ear provides the text. Barthes was perhaps the more willing to describe the photograph as a message without a code because he always imagined the photograph in a context—a newspaper or a magazine, a billboard or a living room wall, a gallery or someone's hand—which would pick the photo up and through a labor of connotation give it the applied meaning without which it would be not a *representation* but just another thing in the world:

> Naturally, even from the perspective of a purely immanent analysis, the structure of the photograph is not an isolated structure; it is in communication with at least one other structure, namely the text—title, caption or article—accompanying every press photograph.[60]

He acknowledged other connotation procedures as well:

> Connotation, the imposition of second meaning on the photographic message proper, is realized at different levels of production of the photograph (choice, technical treatment, framing, lay-out) and represents, finally, a coding of the photographic analogue.[61]

Here he observed its rhetorical exploitation through the use of trick effects, the posing of figures and objects, photogenia, and the effects of aestheticism and the syntax of sequences or collections of photos. Finally he turned to the text:

> The text constitutes a parasitic message designed to connote the image, to "quicken" it with one or more second-order signifieds. In other words, and this is an important historical reversal, the image no longer *illustrates* the words; it is now the words which, structurally, are parasitic on the image.[62]

Is any of this applicable to the Van Sant?

All of it is. We have already seen that, internally, the map exploits every conceivable connotative strategy. Barthes' discussion of *trick photography* revolves around a faked image of then Senator Millard Tydings in conversation with Communist leader Earl Browder, faked in the sense that it was "created by the artificial bringing together of the two faces."[63] "Faked" seems somehow less germane in Van Sant's *artificial bringing together of the two summers of the northern and southern hemispheres*. But in a context stirred up by titles like A *Clear Day* and captions like "First-of-a-kind portrait from space," *trick effect* does not seem too wide of the mark. In his consideration of the effect of the *pose*, Barthes observes the way a particular shot of then President Kennedy—eyes raised, hands together—achieved its effect by virtue of its reference to *a store of stereotyped attitudes*, a historical grammar of iconographical connotations. And this is precisely what the Van Sant does when it portrays the planet *exactly as we have always seen it*, the land bright against the dark blue oceans, its symmetrical disposition around a central Atlantic, Antarctica a band of white across the bottom. Indeed, were it not for its reference to this stereotyped vision of the planet—it *could* have been turned upside down, it *could* have been centered around a pole (it *is* a sphere; *any* disposition of land and water could be displayed)—it could not so effectively have . . . *concluded the history of cartography*. In his treatment of the effects of *objects*, Barthes is above all else concerned with the meaning their selection and composition connote. He "reads" a composition of a vase of flowers with a magnifying glass and photo album before a window opening onto vineyards and tiled roofs: "The connotation somehow 'emerges' from all these signifying units which are nevertheless 'captured' as though the scene were immediate and spontaneous, that is to say, without signification."[64] And is this not how the earth appears in the Van Sant, uncoded, innocent. And yet a connotation emerges from this map that does not emerge from, say, that more snapshot-like image of the earth as a blue–black roundel *alive* with a swirling turbulence of cloud and storm. That earth seems tough, ferocious, powerful. The *suppression* of the clouds in the Van Sant renders an earth so still it seems to be holding its breath, connotes an earth that is . . . *waiting for something*, an earth that is paper thin, that is delicate, fragile.

Barthes saw that the connotation of the magnifying glass and photo album before the window was explicitly unfolded in an accompanying text. Nothing is different in our case:

> Inspiration can spring from adversity. In March 1989 artist Tom Van Sant had eye infections that threatened to blind him. The treatment: antibiotics in both eyes every 30 minutes for ten days. Rather than fight the lack of sleep, Van Sant entered into a state of meditation. At the end of his ordeal he was left with an overpowering sense of purpose—to create a portrait of the earth, derived from satellite imagery, to highlight environmental themes. Purpose became obsession; a year later he completed the map.[65]

This makes everything explicit, explains the map's . . . *devotional* quality, its sense of . . . *veneration*. Suddenly we see that the map is a votive offering, a New Age retablo. And not just for Van Sant, but for Northern Telecom and the National Geographic Society too, each in its way committed to a vision of the earth as sacred, as having an existence . . . apart . . . from that of man.

Here: the dictionary defines a retablo as a votive offering made in the form of a religious picture, typically portraying Christian saints, painted on a panel and hung in a church or chapel (where a votive offering is one made in fulfillment of a vow, often in gratitude for deliverance from distress). Now, here is the English text (it appears in five other languages as well) from *A Clear Day*:

> In December of 1991 Northern Telecom became the first global telecommunications company in the world to eliminate chlorofluorocarbon (CFC-113) solvents from its manufacturing operations. This unique image of the earth was created by Tom Van Sant and the GeoSphere Project by compiling thousands of satellite photographs on a supercomputer to produce the first cloud-free photograph of our planet. Symbolizing Northern Telecom's contribution to the protection of the ozone layer, this poster commemorates that achievement.

One clear day deserves another. The cloud free day that Van Sant created through his arduous sifting of satellite images looks forward to that day when the atmosphere will again be clear of polluting chlorofluorocarbons (that is, when we will be delivered from the distress of ultraviolet radiation). The map not only *symbolizes* the Northern Telecom achievement, it *sanctifies* it, unites Northern Telecom with the earth—or that vision of the earth—embodied in it.

Which is exactly what it does for the National Geographic Society, symbolizes its *concern* for the earth at the same time that it infuses this concern with . . . *the odor of sanctity*. The grammar of iconographical connotations appropriate here must include the cover of the issue with which the *National Geographic* concluded its first century of publication. Headlined, "Can Man Save This Fragile Planet?" it consisted of a hologram of a crystal globe that when tilted was seen to shatter. An interior caption made the point explicitly: "Symbolizing the planet's fragility, a crystal globe appears whole and then shatters as the three-dimensional image is tilted under a single light bulb." Intact versions of the globe were awarded to environmental heroes: the accompanying copy ran, "Our planet, like the delicate crystal globe on its pedestal, deserves the best care we can give it." On the contents page, under the headline, "Will We Mend Our Earth?" we read, "As the National Geographic Society enters its second century, one of its goals will be 'to encourage better stewardship of the planet.' "[66] What kind of a planet? The kind embodied in the Van Sant map, *still, thin, fragile . . . as delicate as crystal.*

When did the earth begin to seem so . . . *vulnerable?* Was it with manned space flights, when, for the first time, we could get high enough for a really *global* perspective? Or was it when we first saw the earth from the moon, first saw it as "spaceship earth," first confronted the terrible isolation of, in the words of The Nature Company, "our frail planet, swirling in space." Whenever it was, the conviction has been growing on many of us that the earth is wounded, delicate, as fragile as the Beach Boys' promise of an endless summer, or Irving Berlin's of "nothing but blue skies, from now on." But it shouldn't take Julian Simon's insistence "that the claims of ecological crisis and dwindling resources are vastly

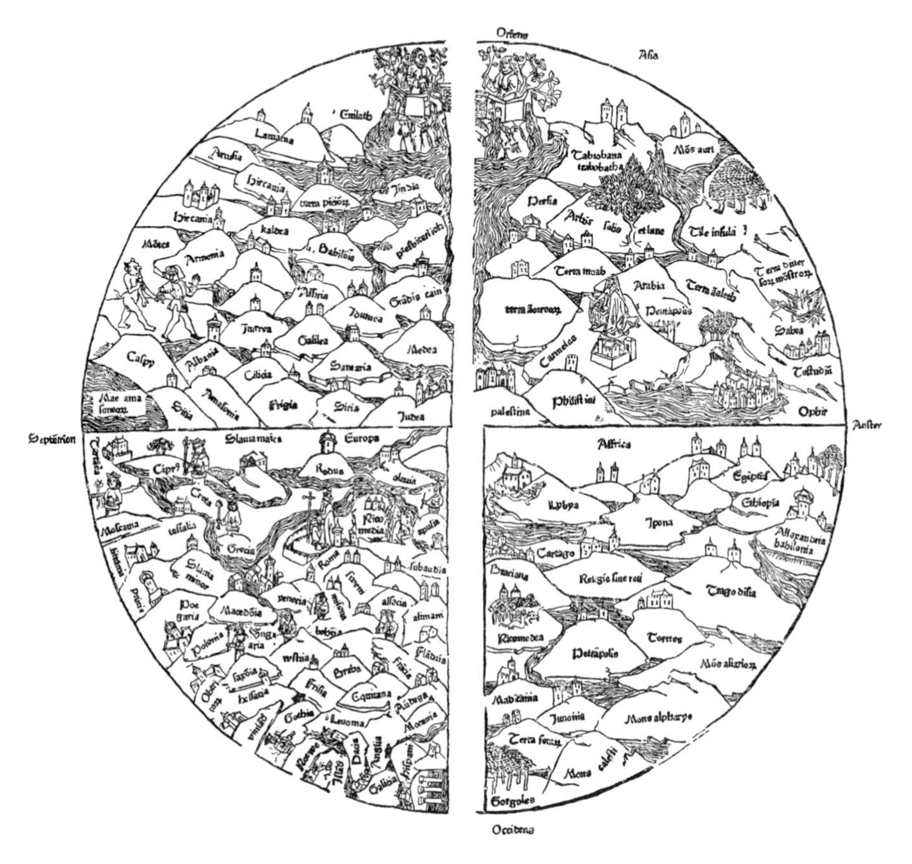

Emblematic of an explicitly *Christian* worldview is this *mappamundi* from *Rudimentum Novitiorum*, Lübeck 1475. It *is* different from Tom Van Sant's map of the world, but not that *much* different.

exaggerated and that 'untrammeled human resourcefulness and enterprise' will forever find new expedients, including new technologies, to offset shortages and ameliorate the environment"[67]; or a description of the Bush administration's attitude toward global warming; or a recitation of the Pope's policy on birth control; to make it clear this view is but one of many. Others see the planet in different terms, remain more convinced of its resilience. In the end the conviction that the planet needs mending is . . . *a religious one*.

In the end, it is hard to distinguish our map from one of those medieval *mappaemundi* that were for so long the touchstone of cartographic backwardness. When David Woodward asks us not to misinterpret the significance of their disk shapes, or to censure out of hand the spread of time they depicted, he could be speaking of the Van

Sant in the *National Geographic*, disk-shaped as it is and containing as it does images collected across a 3-year span of time. But it is when he refers to their function—"primarily to provide a visual narrative of Christian history cast in a geographical framework"[68]—that he most touches our map. If it is not intended to be a narrative of Christian history, the Van Sant *is* intended to be an emblem for a kind of environmental orientation that takes temporal awareness—of how little time is left—as a given. The map is instinct with a kind of millenarian *"Don't drop it. It will break."*

Acme of cartographic perfection though it is, the map thus emerges in the context of a mapmaking society *struggling with its future* to serve an interest, that of those committed to . . . a certain vision of what it means to live. You may share this vision—I do—but it serves *no* interest at all to pretend that it is the planet speaking through the disinterested voice of science, instead of me, Tom Van Sant . . . or you.

CHAPTER FOUR

The Interest the Map Serves Is Masked

"To serve an interest": why when it is made of maps is this assertion so hard to swallow?

We have seen that this interest is no more than the inevitable consequence of being enabled to see what otherwise we couldn't, that it is no more than the unavoidable price we pay for the intrusion of that author who brings this vision into being out of the concrete history of a living, that in the end it is this interest that is what the map offers, that distinguishes the map as *representation* from the world it represents. Why then does this interest—this commitment—of the map . . . *stick in so many craws?*

It is because the map is powerful precisely to the extent that this author . . . *disappears*, for it is only to the extent that this author *escapes notice* that the real world the map struggles to bring into being is enabled to materialize (that is, to be taken *for* the world). As long as the author—and the interest he or she unfailingly embodies—is in plain view, it is hard to overlook him, hard to see around her, to the world described, hard to see it . . . *as the world*. Instead it is seen as no more than a *version* of the world, as a *story* about it, as a *fiction:* no matter how good it is, not something to be taken seriously. As author—and interest—become marginalized (or done away with altogether), the represented world is enabled to . . . *fill our vision*. Soon enough we have forgotten this is a picture someone has arranged for us (chopped and manipulated, selected and coded). Soon enough . . . it is the world, it is real, it is . . . *reality*.

Here, listen to John Van Pelt describe the world of maps in a *Christian Science Monitor* review of Mark Monmonier's *How To Lie With Maps:*

> The primary tool is a healthy skepticism. Any "single map is but one of an indefinitely large number of maps that might be produced . . . from the same data." This learned, we respond with fresh respect to the elegance and exactitude of topographic survey plats, while we energetically question the maps of the proposed housing project, the map depicting concentrations of anything, and that paragon of self-serving cartography, the advertising map.[1]

But how are these maps different from each other? In their embodiment of interest? Not at all. As we have seen, all maps—from the most apparently "objective" to the most blatantly "subjective"—embody the interests of their authors, indeed, *are* the interests of their authors in map form. The maps Van Pelt describes vary only in the degree to which the interest they embody is *patent*, is . . . *rubbed in our face*. The healthy skepticism is brandished only when called upon to defend against *evident* interest (advertising, developers). When the authors have rendered themselves *transparent*, suddenly we have . . . *fresh respect*. And it is astonishing how easily this happens, how readily we take for granted—as natural—what is never more than the social construction of a map. All that is required is the disappearance of the author, the invisibility of the interest.

Here, Van Pelt again. After predictably (but laudably) intoning, "In no case, however, is the map equal to reality," he immediately qualifies himself . . . *on the basis of scale:* "Maps at small scales tend to be less detailed than those at large scales—the mapmaker has less room to illustrate features and hence must be more selective. But maps at large scales also suppress some details."[2] *Some* details? How about . . . *almost all* details? Something in us resists this sweeping assessment, but it is nonetheless true. Outside my window this morning the dogwood across the street is like seafoam. The filtered sunlight on the weathered asphalt makes archipelagos that slip and slide in the breeze. Robins and cardinals, blue jays and doves dive and bank, swoop and holler. I will hear them better once the day warms up and we open the windows, but even with them closed, the planes over head are loud and clear. Joyce walks by with Spot. He lifts his leg . . . and pisses.

I roll out the *Raleigh West Quadrangle/North Carolina—Wake County/7.5 Minute Series (Topographic)* survey sheet. Where a "small" scale map can be as small as 1:100,000,000,[3] its 1:24,000 is big enough to qualify as "large," but I scour the map in vain for these details. Under the red tint that covers my neighborhood only landmark buildings are shown, landmark buildings, hydrography, contour lines, and names. I can see well enough where the house *should* stand, directly under the second "a" in "Cabarrus." There is the road all right, but not the shifting shade or even the cars that despite the early Sunday morning hour do not hesitate to hustle down the road. The house *is* there on the larger scale insurance maps that Sanborn used to make: at 1:600, these give its footprint and distinguish porch and address. They even show the garage we took down the day we moved in. But they don't note the squirrels or the pecan tree they nest in, though this is larger (and in the long run more important) than the house. At much larger scales, the maps turn into plans: on a bunch done by landscape architecture students, the pecan tree does make an appearance. But now Cabarrus has disappeared and with it the dogwood across the street.

"Irrelevant," one mutters: "No one insures *trees*." "Too ephemeral, too transitory," says another. "Too auditory," pipes up a third . . . but what is revealed thereby is the *prior* editing the world endures, the way *before we get to the map* the domain of expected detail has been throttled down to a certain class of objects. "Birds and bees? The mapmakers weren't *interested* in those things." Exactly. So what did they map? What they *were* interested in. And this is the interest the map embodies . . . inevitably.[4]

This is easier to see in some maps than others. The interest all but boils off *McCormick's Map of the World*, bubbling as it does with the company name (49 occurrences), its products (illustrations of 13 bottles, jars and boxes), logo (in the center of the windrose) and coat of arms (with its motto wrapped around the globe). Among vignettes illustrating the harvesting of coffee, tea, vanilla and spices are those displaying McCormick's home office in Baltimore, its southern California headquarters in Los Angeles and its Schilling Division offices in San Francisco. Further vignettes of famous explorers bracket another—top and center—of the American flag over Fort McHenry (Baltimore, site of McCormick's home office) triumphantly waving—by implication of its position—*over the world* . . . the flag's of whose other nations ("Flag Research Courtesy U.S. Navy Hydrographic Office") form the map's colorful border. Countries are named, with spice sources indicated (thus: Siam, "White Pepper"), an inset labels the Moluccas "The Spice Islands," the lavish text comprises "McCormick's Encyclopedia of Flavoring Extracts–Spices–Teas–Coffees." What does the map say? It says, "McCormick: From all the World, Known the World over." It says, "Buy McCormick."[5]

Here's another map feverish with interest, *San Antonio*, a chili con carne ("originated in S.A. in the 1840's and served to the public by chili queens") of people, places and things, spiced with hints of road, Riverwalk and promotional copy ("Schilo's Delicatessen," "America's first public housing for the elderly was in San Antonio," "Lone Star Beer," "Brook's A.F.B. est'd 1917," "Ingram Park Mall"). In the lower corners large vignettes refer to history (colonists, pioneers) and Mexico (mariachi band). Jumbled, crowded, *hot*, the map says, "Historic, modern, bustling, exciting, Mexican-flavored San Antonio!" It says, "When you get back home tell everyone how much fun you had here!" It says (all but explicitly), "Build your next plant . . . in San Antonio."[6]

A final example, *Delta's Domestic Route Map*, that is, the United States, and parts of Canada, Mexico and the Caribbean . . . all but obscured beneath a thick weave of blue lines symbolizing not merely Delta's *routes*, but the *embarrassing abundance* of Delta's routes. What does the map says? It says, "We blanket America," that is, "We will keep you so warm you will never want to get in bed with another carrier." Does the map lie? No, why should it bother? Delta has no interest in selling you a ticket on a route it doesn't fly. The point is merely to dissuade you—through the exploitation of age-old rhetorical devices (emphasis, exaggeration, suppression, metaphor)—from thinking of American or TWA or USAir next time you want to fly.

But here's another map, another United States Department of the Interior Geological Survey sheet, the *Wanaque Quadrangle*, that is, the Wanaque Reservoir and parts of neighboring hills and towns . . . all but obscured beneath a thick weave of contour lines symbolizing not merely Wanaque *topography*, but the *embarrassing abundance* of Wanaque topography. What does the map say? It says, well it . . . *What does it say?* In the first place it's not nearly as univocal as *Delta's Domestic Route Map*. Where that map makes its point and then moves in for the sale, this map . . . *babbles*. Not content to talk about the topography, it goes on to discuss the extension of urban areas between 1954 and 1971. It mutters about surface waters. Every other phrase is about trees. Yet its manner is disarming: it seems so uncertain what it wants to say that it seems hard to imagine it wants to say anything, or that it wants to say anything . . . *in particular*. What's its point? It would seem, in the end, to have none.

Which *is* its point. Where have we seen this before? *In almost every map we have looked at*. Again and again we have seen a similar *vagueness* of content and form, a similar *diffusion* of ends and means. The map will show everything (and therefore claim innocence about the choice of anything) and will show it as it is (ignoring the "white lies" the map must tell in order to be accurate and truthful). The map will be seen to serve so many purposes that none can predominate, or its means will be so widely spread in so many social institutions that it can be claimed by none. Responsibility for the map will be shuffled off onto layered and competing user groups, or its authorship will be shown to be fragmented among so many specialists as to be impossible to establish. Lying and lost, vague and confused, the map will *therefore* show the world the way it really is. John Garver praised the Robinson projection for precisely these reasons: ". . . we believe that its compromises are the most reasonable for a general reference map of the world"—despite the exaggeration of the size of Russia and the United States, the diminution of Africa, the

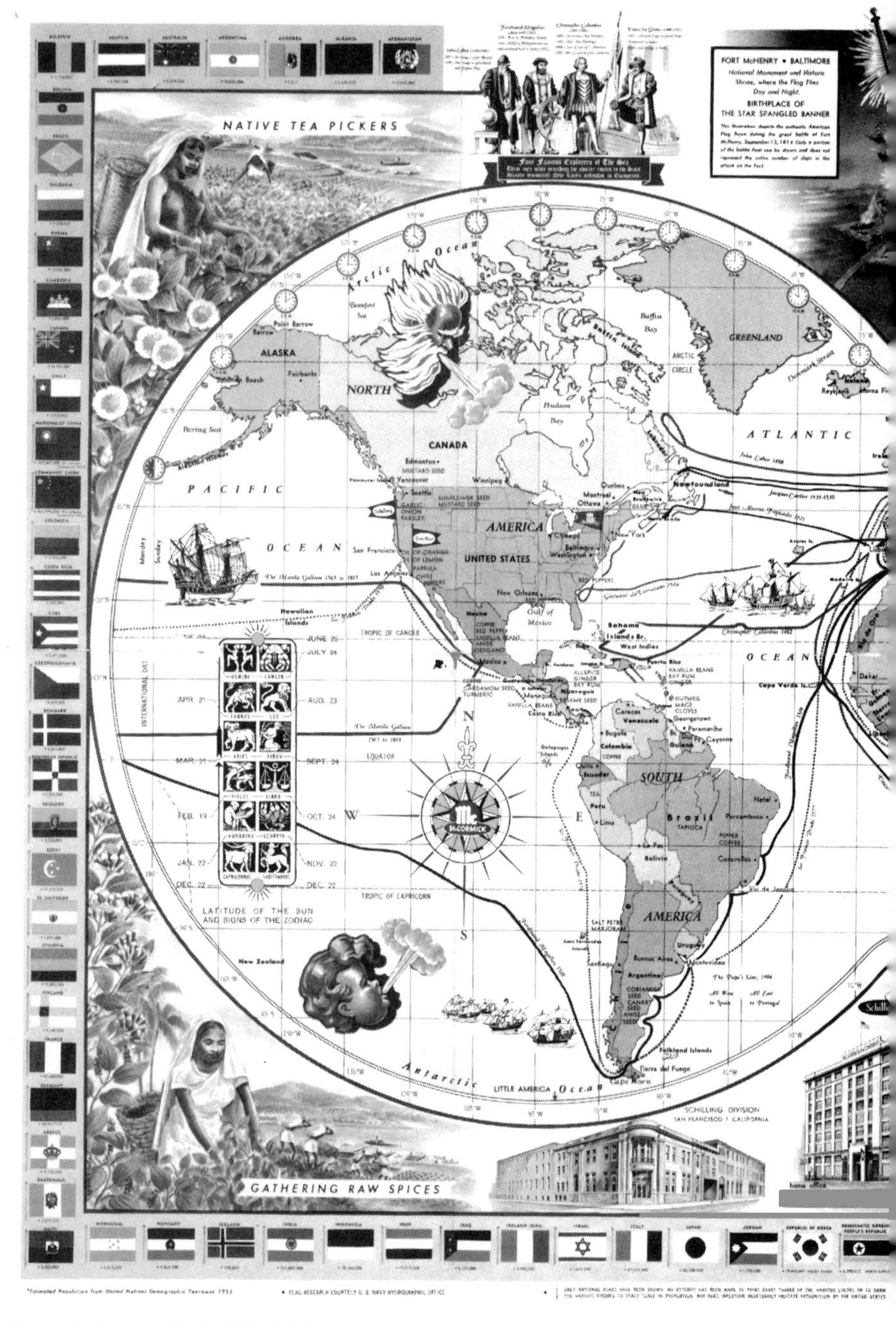

"McCormick: From all the World, Known the World over." (McCormick's *Map of the World* is no longer published and copies are not available.)

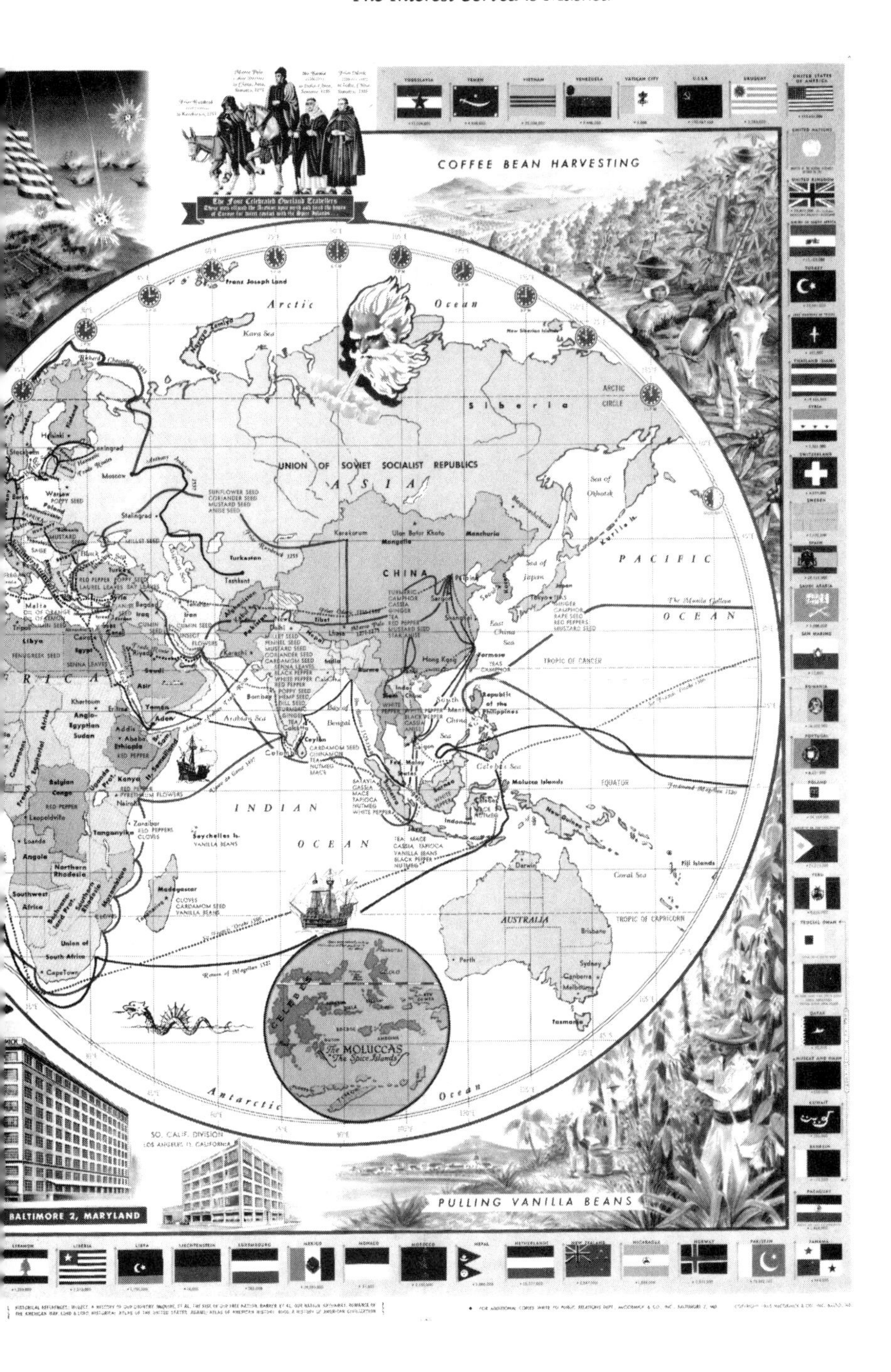
COFFEE BEAN HARVESTING
PULLING VANILLA BEANS
BALTIMORE 2, MARYLAND
SO. CALIF. DIVISION
Arctic
Ocean
PACIFIC
OCEAN
INDIAN
OCEAN
Antarctic
Ocean
UNION OF SOVIET SOCIALIST REPUBLICS
ASIA
Siberia
CHINA
AUSTRALIA
The MOLUCCAS
The Spice Islands
TROPIC OF CANCER
TROPIC OF CAPRICORN
EQUATOR
Moluccas Islands
Bay of Bengal
Arabian Sea
Sea of Japan
Sea of Okhotsk
Celebes Sea
Coral Sea
Madagascar
Ceylon
Seychelles Is.
Fiji Islands

"somewhat compressed" appearance of Greenland *et cetera et cetera* —precisely because it results in this highly desirable kind of . . . *pointlessness:* "The projection does not espouse any special point of view."[7]

The Naturalization of the Cultural

What is so desirable about this pointlessness achieved with such an effort of willful misconception and delusion, of self-deception and denial? It is the ultimate alibi, it is that most effective mask, it is . . . *the world.* What is the *point* of the world? *It doesn't have any,* it is pointless. The interest unavoidably embodied in the map is thus disguised . . . as natural; it is passed off as . . . Nature itself. Only the little white lies *required by the world* get in the way.[8] As Mark Monmonier puts it:

> Not only is it easy to lie with maps, it's essential. To portray meaningful relationships for a complex, three-dimensional world on a flat sheet of paper or a video screen, a map must distort reality. As a scale model, the map must use symbols that almost always are proportionally much bigger or thicker than the features they represent. To avoid hiding critical information in a fog of detail, the map must offer a selective, incomplete view of reality. There's no escape from the cartographic paradox: to present a useful and truthful picture, an accurate map must tell white lies.[9]

Notice how the sunlight on the street, the birds singing in the trees, the shining dogwood turn into a . . . *fog,* the way exaggerating the size of Russia and the United States at the expense of Africa is . . . *essential,* the way the confusion of truth and lies is passed off as . . . *a technical problem.* In fact they're all passed off as technical problems (not as questions of philosophy), the map *must* offer a selective, incomplete view of reality (*so don't blame me*), and the end result is that except for a little *cartographic mischief*—political propaganda, military disinformation, advertising maps (*McCormick's Map of the World*)—the map remains . . . *pointless,* and therefore the Nature it displays remains essentially . . . *intact.*

What response should be mounted to the bland front of this Naturalization of the Cultural? For Roland Barthes:

> The starting point of these reflections was usually a feeling of impatience at the sight of the "naturalness" with which newspapers, art and common sense constantly dress up a reality which, even though it is the one we live in, is undoubtedly determined by history. In short, in the account given of our contemporary circumstances, I resented seeing Nature and History confused at every turn, and I wanted to track down,

> in the decorative display of *what-goes-without-saying*, the ideological abuse which, in my view, is hidden there.[10]

It is precisely the *what-goes-without-saying* quality of the classical cartographic defense of its practices that is most insidious, not only because it renders the inevitable interest invisible to those who view the map, and who have been induced by a profound cultural labor to accept it as the territory, but because it renders this interest invisible to the cartographers as well who manage in this way to turn themselves into . . . *victims of the map*.

What this amounts to is a form of repression that is, not just the denial of an interest or a point of view (let's face it, God alone has no special point of view), but a denial that anything was denied. R. D. Laing puts it this way:

> When I was thirteen, I had a very embarrassing experience. I shall not embarrass you by recounting it. About two minutes after it happened, I caught myself in the process of putting it out of my mind. I had already more than half forgotten it. To be more precise, I was in the process of sealing off the whole operation by forgetting that I had forgotten it. How many times I had done this before I cannot say. It may have been many times because I cannot remember many embarrassing experiences before that one, and I have no memory of such an *act* of forgetting. I was forgetting before I was thirteen. I am sure this was not the first time I had done that trick, and not the last, but most of these occasions, so I believe, are still so effectively repressed that I have still forgotten that I have forgotten them. This is repression. It is not a simple operation. We forget something. And forget that we have forgotten it. As far as we are subsequently concerned, there is nothing we have forgotten.[11]

It is easy but disingenuous to froth at the mouth over the flimflams of advertisers, or the falsehoods of Nazi propagandists, or the charmless fabrications of military disinformation specialists.[12] But perhaps disingenuous is too kind a way to put what in the end is no different from all these other half-truths, for the fact is all these more or less self-evident falsehoods owe whatever authority they have to the great national mapping projects, to the wall maps in the thousand classrooms, to the road maps the gas companies used to give away, to the plates in atlases, the cuts and insets in newsweeklies and textbooks, each of them, as Monmonier admits, built from *white lies* about which these maps are all but *entirely* silent. Because it is the cartographer and not the graphic artist or political cartoonist who has first repressed the magnitude and significance of his intervention in what passes in the map for a transcription of nature, it is in precisely the *cartographer's products* that the repressed experience—the interest represented, the point of the map—must be sought. This is not only because it is essential to understand what this interest *is*, but because it is its *repression* that enables the map to masquerade so effectively as truthful and accurate, that is, to provide the necessary context for the "propaganda," advertising and military disinformation maps to work. Thus the problem was never "*how did this map fool me?*" but "*why was I so inclined to wholeheartedly believe in it in the first place?*" From this perspective:

> Maps cease to be understood primarily as inert records of morphological landscapes or passive reflections of the world of objects, but are regarded as refracted images contributing to dialogue in a socially constructed world. We thus move the reading of maps away from the canons of traditional cartographic criticism with its string of binary oppositions between maps that are "true and false," "accurate and inaccurate," "objective and subjective," "literal and symbolic" or that are based on "scientific integrity" as opposed to "ideological distortion." Maps are never value-free images; except in the narrowest Euclidean sense they are not in themselves either true or false. Both in the selectivity of their content and in their signs and styles of representation, maps are a way of conceiving, articulating and structuring the human world which is biased towards, promoted by, and exerts influence upon particular sets of social relations. By accepting such premises it becomes easier to see how appropriate they are to manipulation by the powerful in society.[13]

This said, Harley turns to an historical survey (he was, of course, an historian). But because this has the effect of isolating such practice in the past ("*I used to be bad, but I'm not anymore!*"), it is essential to stress the extent to which such manipulation is an aspect of neither this nor that historical period, this or that society, but rather a property . . . *inherent in the map*.

The Culturalization of Natural

Before the cultural content of the map can be naturalized *out* of existence, the natural content of the landscape must be culturalized *into* existence.[14] This is a labor of culture, it is a labor of identifying, of bounding, of naming, of inventorying . . . it is a labor of mapping. Since these processes occur bundled up together in the living that human occupance of the land amounts to, there is no first place from which to launch ourselves. The land is not *systematically* divided into plains and hills that are subsequently delimited, named and inventoried. Instead, the human landscape is brought into being historically, sometimes in a rush, but usually bit by bit in a patient dance of disjointed incrementalism. As we know, the map is not an innocent witness in this labor of occupance, silently recording what would otherwise take place without it, but a committed participant, as often as not driving the very acts of identifying and naming, bounding and inventorying it pretends to no more than observe. To sketch a river is to bring into being—inescapably—the land that it drains; what was originally whole is suddenly in pieces—water, banks, slopes, hills—which, as they materialize take their places (if only vis-à-vis each other) and soon enough, names. This is, of course, a human way of being in the world—mapping is a way of making experience of the environment *shareable*[15]—but in map*making* societies these activities take on a number of less explicit functions:

> The intentional meaning behind the application of the distinctive federal public-land-survey grid to the official topographic series published by the U. S. Geological Survey is straightforward: to assist in locating areas and to assign exclusive coordinates to them. The implicational meaning lies elsewhere in the related concepts of resource inventory, identification, allocation, and purchase of private property; property protection and access through thousands of miles of barbed-wire fencing and pavement; manifest destiny; and the geometry of American society. The act of designing and producing such a map is an action of subjugation and appropriation of nature, a basic value of American society, not merely the reification of an idle curiosity in recording dimensions.[16]

It is precisely to the extent that the map culturalizes the natural that the cultural production the map *is* must be naturalized in turn, this to make it easier to accept—*as natural*—the historically contingent landscape the society that wields the map has brought into being.[17] But if we are to understand how this process actually *works*, that is, how maps *accomplish* this tremendous labor, we must turn to the map, and not one from a distant time or place, or one marginalized by its origins within our society, but one from the heart, one from the core . . .

The Wanaque Topographic Quadrangle . . .

"United States/Department of the Interior/Geological Survey" it says in the upper left, "Wanaque Quadrangle/New Jersey/7.5 Minute Series (Topographic)" in the upper right. In between it says, "United States/Department of the Army/Corps of Engineers." The multiple authorship implied here is clarified in a block of type in the lower left. "Mapped by the Army Map Service," it says there, "Edited and published by the Geological Survey." On the next line it adds, "Control by USGS, USC&GS, and New Jersey Geodetic Survey." One is all but over-

The United States Geological Survey's *Wanaque Quadrangle, New Jersey, 7.5 Minute Series (Topographic).*

whelmed by the implications of all this bureaucracy. This *thin sheet of paper* subsumes—and brings forward for potential action—labor performed by four divisions of two departments of the federal government and one of a state government. *You wouldn't imagine it to look at it.* The hills and the streams, the lakes and the towns, the all but unbroken cover of the Norvin Green State Forest, seem unburdened by the pressure of even a single governmental agency. Though attributions to the Army or to the Department of the Interior affect every reading of any image they frame—ineluctably they construct a certain kind of faith (while they undermine others)—the Corps of Engineers, the Geologic Survey, the New Jersey Geodetic Survey, seem unrelated to the content of the map, which, despite everything, appears here to be no more than . . . *observed:* a sewage disposal plant, Oakland, a cemetery, Wolf Den Dam . . .

. . . Shows Only Selected Features

The heterogeneity, like the diffusion of authorship through so many layers of government, speaks for the . . . impartiality . . . of the world described here: a place for everything, everything in its place:

> Unlike special-purpose maps, the quadrangle maps produced by the Survey are designed to be used for many purposes. Scales, contour intervals, accuracy specifications, and features that are shown on the maps have been developed gradually to satisfy the requirements of government agencies, industry, and the general public. Because these maps serve a wide variety of uses, they are called general-purpose maps. The functions a map is intended to serve determine which features should be mapped, *but other factors are taken into account before it is decided which features actually can be shown.*[18]

This, from a sort of official survey of its products published by the U. S. Geological Survey on the occasion of its hundredth birthday, goes on to list three such factors: the *permanence* of the features, the *cost* of compiling the information, and map *legibility.*

The Wanaque Quadrangle Only Shows "Permanent" Features

Permanence of the features? What could this mean? "Purple indicates the extension of urban areas," it says on our map, and, "revisions shown in purple complied by the Geological Survey from aerial photographs taken in 1971." Because none of the revisions was field checked—which makes it sound like the map was *revised* instead of *updated* to grab a landscape in flux—they aren't named. But there are what? *Over fifty new ponds? A dozen new subdivisions? Miles of new road?* Maybe I'm missing something, but where's the permanence in this? What we sense here is . . . transition, change.[19] What we also sense is the presence of a code: permanent = important. Needless to say, this is not made terribly explicit, on or off the map: "Cultural features are especially subject to change. If the maps are to have a reasonably long useful life, the features portrayed must be restricted, to some extent, to relatively permanent objects" is all Thompson has to say about it[20]; and though this may explain why the birds in my backyard don't show up on the *West Raleigh Quadrangle*, the remark is otherwise of the surpassing vagueness we have come to recognize as an attribute of the . . . *objective*. Another way to approach this is to look at the "topographic map symbols" the Survey uses. This *legend*—which is not printed on any of the quadrangles (because the symbols are presumed to speak for themselves?) constitutes a *catalogue* of the features the Survey deems permanent enough to map. In 1979 there were 106 of these: *primary road, hard surface; secondary highway, hard surface; light duty road, hard or improved surface; unimproved road; trail; railroad, single track; railroad multiple track; bridge; drawbridge; tunnel;*

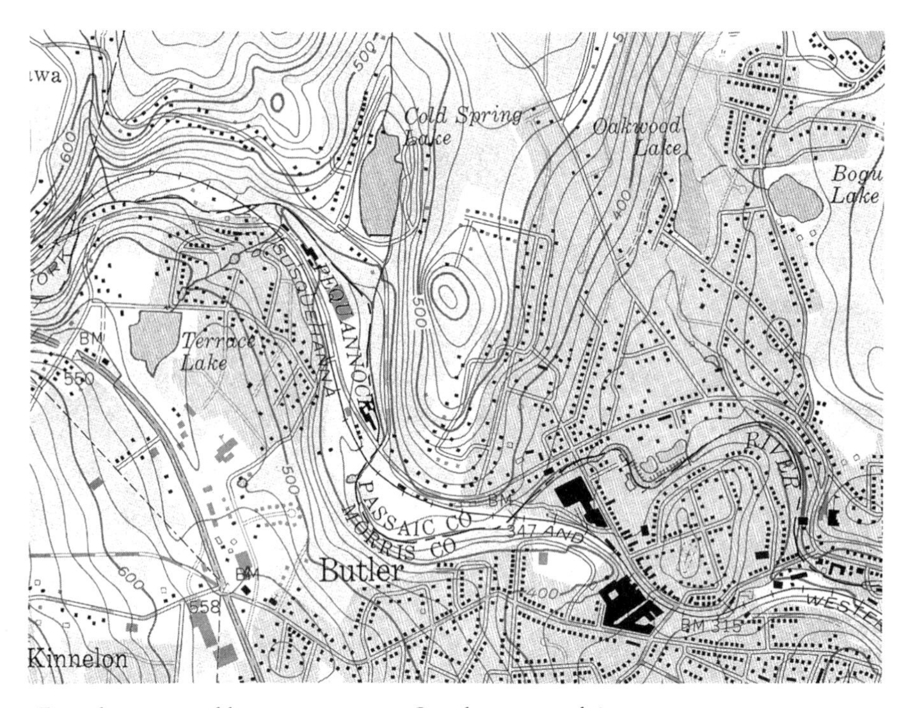

Everything mapped here is permanent. On what time scale?

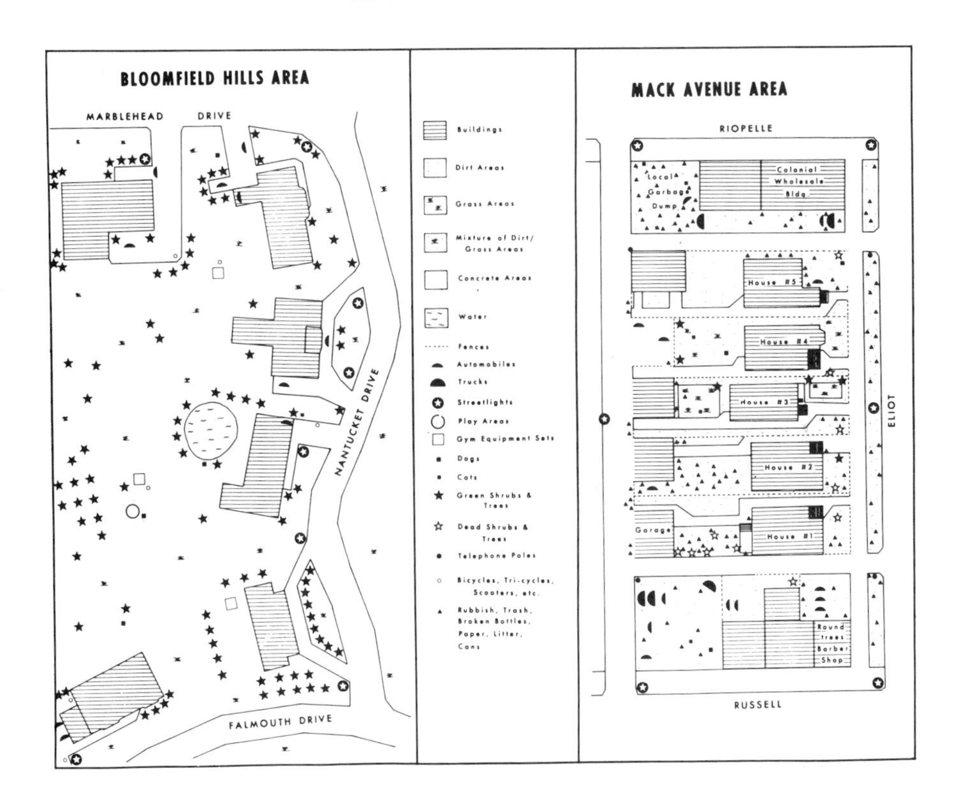

Mapped here from the Bloomfield Hills and the Mack Avenue areas of Detroit are things the U.S. Geological Survey overlooks: dead shrubs and trees, broken bottles, paper, litter, bicycles, tricycles, scooters . . . (From the Detroit Geographical Expedition and Institute, *Field Notes: Discussion Paper No. 3, The Geography of the Children of Detroit*, Detroit, 1971.)

footbridge; overpass; underpass . . . You get the idea. Permanent features included *campsites, picnic areas, exposed wrecks, fence or field lines, glaciers, intermittent streams and land subject to controlled inundation.*[21] Though this list hardly clarifies what is meant by "permanence of the features" (it is littered with so many contradictions it seems churlish to add that the definition of "permanent" is, *au fond* . . . political), there is nevertheless so sturdy a commonsensicality about it all that doubt is all but put to sleep: what else should the Survey *map?*

It all depends on the intention. At the moment I am looking at two other maps, one of a block in Bloomfield Hills, the other of a block in the Mack Avenue area of Detroit.[22] Here are some of the things on *their* common legend: *automobiles; trucks; dogs; cats; green shrubs and trees; dead shrubs and trees; bicycles, tricycles, scooters etc.; rubbish, trash, broken bottles, paper, litter, cans.* Bloomfield Hills is clotted with grass, green shrubs, and trees. Bicycles and tricycles are everywhere. In the yards, which are unfenced, are three gym sets, a play area and a pond. In the Mack Avenue area, dead shrubs and trees outnumber the living, there is more litter than grass. In the yards, most of which are fenced, are no bicycles, gym sets, play areas or ponds. Evidently the two places are markedly different, but it is not a difference the Geological Survey acknowledges. Why? Because few of the things that mark it are mapped by the USGS. The reason? *Because they're not permanent*. Of course they *are*, the grass and toys of Bloomfield Hills are as permanent features in Detroit as the trash and broken bottles of Mack Avenue, *more* permanent, in fact, than the buildings that went up in the flames of 1967 or in the slower conflagration that followed.[23] But permanence never had anything to do with it: the Survey has no interest in the durative. That it's not an issue is written all over the Survey legend, *all over the quadrangles;* the issue lies elsewhere, in property perhaps (and 13 different boundaries are mapped on topo quads, to say nothing of section corners and boundary monuments), or in concrete (most of the culture of the Survey maps is cast in concrete), or in some other function better served by whatever agenda it is the Survey ends up actually following, which (let us admit it now) does not have on it anywhere a concern for the relative hostility of the American landscape to children.

The Wanaque Quadrangle Only Shows Cheap Features

In this regard a topo sheet is deceptive: the apparent richness of the detail promises a *deeper* understanding than can be struck. Take this factory in the floodplain of the Pequannock River, for example, way down in the southeast corner of our quad, all but off the map, an industrial intruder in this landscape of blue waters and green hills. The fine lines behind it speak of spurs from the New York, Susquehanna and Western railroad. A light road crosses them, runs back into the lot behind the building. I'll bet old C-clamps and spike plates are rusting there along the right-of-way where the goldenrod and Queen Anne's lace are running riot. In the scrub beside the road, old pallets and cable spools are probably rotting, and in the mornings when the mist lies just above the river you can probably *taste* the creosote on that heavy bottom air. Or maybe not. Maybe the lot is paved from the tracks to the river, maybe brightly colored 50-gallon drums are stacked 10 deep inside a chain-link fence festooned with concertina wire. Maybe the smell is so lethal all you want to do is get away. But these are not details the quad can spare, the purple tint covers all, and one wonders: did the survey silently delete features present on the 1954 edition but absent from the airphotos taken in 1971? Or were they . . . *left behind* . . . to give some substance to the idea of permanence?

Whatever was done was done with the approval of the Office of Budget and Management: "The extent to which some kinds of map features are shown is determined partly by the cost of compiling the information."[24] How different this sounds from Monmonier's, "to avoid hiding critical information in a fog of detail, the map must offer a selective, incomplete view of reality."[25] Monmonier's constraint derives from an inherent "paradox" in cartography, that is, from one "natural" to mapping. The survey is more candid; it's a question of budgetary—that is *political*[26]—priorities:

> Aerial photographs are the source of most map information, but features that cannot be identified on photographs must be mapped by field methods, an expensive procedure. As an example, not all section corners are shown; they are too small to be seen on aerial photographs, and the cost of mapping them using field surveys would be excessive.[27]

But what this means *in effect* is that nothing that can't be photographed is mapped. Like what? Well, like what the factory does, the number of people who work there, how much they get paid, how much they're worth, how much is shipped out on the New York, Susquehanna and Western, what the land is worth, how much it's taxed, where the folks who own it live, the kinds of wastes they dump in the Pequannock, the way the factory smells, its *sound* . . .

Cheap Maps Are Silent

It's easy to scoff at this, but those who do haven't tried to sleep near a drop forge, with beds *blocks away* rising from the floor with each crash of the hammer—*boom, boom, boom*—every 2 minutes, night after night. Or a speedway. It's easy for Sherre Glover to say:

> In response to the May 7 article "Homeowners raise a din about speedway noise": Oh, come on now. Have we come to the point where we can no longer tolerate any inconvenience even for a few hours one night a week while a majority of people are enjoying themselves? Why don't people check for things such as speedways, airports and hog parlors before moving into a new area . . .[28]

But in a world where sounds and smells are rarely accorded the status of things we can photograph, it's not as easy to do as to say. Certainly it's cheaper not to map them, given the way our national mapping program has evolved, but this was never independent of the priorities and prejudices brought to bear on it. Yet as Bunge and Bordessa have observed, "sounds can tell an extraordinary amount about an area."[29]

Kevin Lynch made similar observations as early as the first edition of his site planning text,[30] and the idea is central to his *Managing the Sense of a Region*, which deals with "what one can see, and how it feels underfoot, and the smell of the air, and the sounds of bells and motorcycles, how patterns of these sensations make up the quality of places, and how that quality affects our immediate well-being, our actions, our feelings and our understandings"[31]—that is, which deals *with a more comprehensive reality* than can be caught in a photo.[32] In *The Tuning of the World*, R. Murray Schafer torqued the idea into "a theory of soundscape design."[33]

Each of these projects demonstrated the plausibility of soundscape mapping. Bunge and Bordessa mapped the quiet groves in Christie Pits: "In Christie Pits, the only quiet groves are the church yards on Ossington, and Leeds Ave., the back lanes, the schoolyards on a non-school day, and parts of the park. Any corner of this community where one can have a quiet game of marbles or where there is a bench to read a newspaper is a sought after commodity."[34] Lynch reproduced a map Michael Southworth had created in an exploratory effort "toward escaping the visual bondage of the contemporary city."[35] Although Lynch found Southworth's soundscape of central Boston *too detailed* for regional analysis, he nonetheless advocated small scale "mapping of the audible field of selected desirable or undesirable sounds (church bells, music, birdsong, early morning garbage cans, helicopters, jackhammers and so on)."[36] Schafer reproduced an isobel map of Stanley Park in Vancouver as well as a "sound map" compiled from "listening walks." He also explicitly addressed the conflict between visual and acoustic space in terms of property:

> A property-owner is permitted by law to restrict entry to his private garden or bedroom. What rights does he have to resist the sonic intruder? For instance, without expanding its physical premises, an airport may show a dramatically enlarged noise profile over the years, reaching out to dominate more and more of the acoustic space of the community. Present law does nothing to solve these problems. At the moment a man may own the ground only; he has no claim on the environment a meter above it and his chances of winning a case to protect it are slender. What is needed is a reassertion of the importance, both socially and ultimately legally, of acoustic space as a different but equally important means of measurement.[37]

Is the Geological Survey remiss in not mapping sounds? Not necessarily. No map can show everything. Could it, it would . . . *no more than reproduce the world*, which, without the map . . . *we already have*. It is only its *selection* from the world's overwhelming richness that justifies the map; it is only its selectivity, its attention, its focus on this at the expense of that, its enthusiasm, yes, its *passion*, that distinguishes the map from the world it represents. It is only *because* it doesn't show everything—or anywhere near everything—that the map has *any* claim on our attention. It is not that every map must be all inclusive, but that maps must come clean about, *must face up to*, the embodied interest that drives—and energizes—their selectivity, that is, their historically driven contingency. Behind the bland face of *permanence* and *cost* lie the real interests the map serves, but, repressed, the map not only denies their existence, it denies that it's denying them.

Here, Thompson again, discoursing on the selection of mappable features, but this time with respect to special-purpose topographic maps:

> For example, a map made for the purpose of designing a new highway would show the type of woodland cover and the classification of soil and rock along the route. Information about drainage, property lines, and buildings would be shown in detail as required. The map would be in the shape of a strip and would cover a relatively small ground area.[38]

It sounds like: *what else would you need?* From the perspective of a highway engineer, probably nothing. But how comprehensive is it really? How wide a range of interests does it actually embody? It depends . . . *on your point of listening*. Here, this paragraph from a local paper:

> More than 150 North Raleigh residents gave state transportation officials an earful Thursday on plans to widen a 9.7-mile stretch of the Beltline—and more than double the traffic noise—near their homes. "If anybody from state government doesn't know what noise is, you are all invited to my house for a backyard party," Sylvia Ruby of Stanley Drive said at the Highway Building auditorium. "You won't hear a thing that's said in the backyard."[39]

But it's not that state and other officials aren't aware of the problem, don't measure noise, even map it; it's that noise fails to achieve that *taken-for-granted* quality of features mapped by the topographic survey, is marginalized, often as *nonphysical*, as though sound were not subject to the laws of physics, could not make life miserable, could not bring about a ringing in the ears, could not cause death.[40] It is this *isolation* of everything *not* on the map that so potently *naturalizes* what's *on* it (what's not on the map . . . *isn't real*).

Legible Features on the Wanaque Quadrangle

A third filter. Having run the world through that of the permanent, and then—under the name of cost—through that of the visible, we will now drip it through the charcoal bed of the . . . *legible*. This is Thompson, but he is echoing academic cartographers everywhere:

> The legibility requirement means that small features must be represented by symbols that are larger than true scale. For example, roads are shown at least 90 feet wide on 1:62,500-scale maps despite the fact that they are actually narrower. Buildings and other structures also are depicted by symbols that may be larger than the scale size of the features. If smaller objects were represented at their true scale size, the symbols would be too small to be legible. Symbols larger than scale size take up extra map space; therefore, where small features are close together, the less important features are omitted in congested areas.[41]

It's wonderful the way this is all so logical: this requires that, and that requires this and *therefore* . . . It's not that we *want* to omit anything (except less permanent features or those too expensive to map), but that we're *compelled* to. Logic and the physical limitations of the eye . . . *insist*. Except for the intrusion of that phrase "less important" the matter seems almost too technical (and therefore too trivial) to even mention. But here's Eduard Imhof:

> The topographical map shows more than a photograph. It is not only a metrically and graphically produced ground plan of the earth's surface, but should also present a wide variety of information which could not be picked up from a direct image such as an aerial photograph.[42] Due to scale restrictions, the cartographer makes a selection, classifies, standardizes; he undertakes intellectual and graphical simplifications and combinations; he emphasizes, enlarges, subdues or suppresses visual phenomena according to their significance to the map. In short, he *generalizes*, *standardizes*, and *makes selections* and he recognizes the many elements which interfere with one another, lie in opposition and overlap, thus *co-ordinating* the content to clarify the geographical patterns of the region.[43]

Coming from a cartographer, the language is extraordinary: *intellectual simplifications and combinations*, *emphasizes*, *subdues*, *suppresses* . . . Imhof's cartographer is far from the automaton "objectively" omitting "less important features" that Thompson evokes. Instead, he is a scientist who, in clarifying geographical patterns, *creates knowledge*, knowledge that by definition is *instrumental*.[44] In such a context, legibility is less a matter for the eye than the mind, and this brings to the surface the question that Thompson elides—along with the rest of those laboring to keep cartography in its psychophysical dungeon[45]—and that is . . . make *what* legible? What is it that with all this machinery of decision making—what to show, how to show it—we are laboring to see?

What Are We Looking for in New Jersey?

"Ringwood": hard to miss this word in the upper right of the Wanaque Quadrangle, one of only four set in 12 point type, "RINGWOOD," "WANAQUE," "WEST MILFORD" "OAKLAND." To what do these names, the most prominent on the map, refer? They refer to boroughs of New Jersey (three in Passaic County, one in Bergen), a borough being the municipal corporation proper to New Jersey, that is, a town or a village. Ringwood village proper doesn't actually appear on the *Wanaque Quadrangle* but on a neighboring quadrangle, *Greenwood Lake*. As the *Wanaque* does, so *Greenwood Lake* too speaks of an embarrassing abundance of topography. It presents the country as a green skin slashed by a long pond (and in fact Greenwood Lake was once known as Long Pond) liberally pocked with warts and knots and welts and blisters, with hills, that is, hills and mountains . . . and lakes and streams, *completely wooded*—indeed much of it lies within the Abram S. Hewitt State Forest—except in the northwest where orchards slip down the northwest faces of Round Hill and Taylor and Warwick Mountains. The startling presence of the Appalachian Trail reminds us that among the uses for topographic maps which Thompson lists (hunting, fishing skiing, camping . . .), hiking figures second.[46] The copy of this quadrangle on the table before me was not photorevised in 1971 like the *Wanaque*—though like the *Wanaque* it was created in 1942 with culture revised in 1954—and so in butting them together I'm slamming 1954 into 1971. This explains why they look so different, the purple tint screaming "change" absent from the *Greenwood Lake*, which therefore looks older, looks less modern, less . . . *developed*. "Ringwood" is a name that appears often on the *Greenwood Lake*, in 12-point type in the borough name, in 10-point type over the cluster of houses that comprise the village, in 6-point type in "Ringwood Creek," "Ringwood Mill Pond," "Ringwood Manor State Park," and in even smaller type along "Ringwood Avenue" running out to Hewitt. But, *whoa!* What are these crossed picks all over the place, *look*, they're, they're . . . *iron mines? Inactive iron mines?*

We're Looking for Iron

In New Jersey? Absolutely. "Ringwood," Allan Nevins reminds us, "might well be called the birthplace of the American iron industry."[47] A settler under George II was the first to build a forge along the Ringwood River, he points out; not much later the Ogden family formed the Ringwood Company, and in 1742, it put up its first furnace. At the time iron was smelted with charcoal—even a small furnace consumed a thousand acres of forest a year—and waterpower was required both for the "blast" and the working of the resulting metal.[48] Little surprise then to find the Ogdens supplementing their purchase of the few acres around the original forge with enough land to give them control of most of the Pequannock and Wanaque Rivers. A little after the French and Indian wars, Peter Hasenclever acquired the property for the London Company, adding to it 10,000 acres around Ringwood and Long Pond acquired from the colonial government. Hasenclever had an 860-foot dam built across the lower end of Taxito Pond—now Tuxedo Pond (on the neighboring *Sloatsburg Quadrangle*)—to provide the waterpower needed to boost the Ringwood works to a capacity of 20 tons a week. Under the subsequent management of Robert Erskine, the mining operation churned out iron products for Washington's Continental Army. Of course the Ogdens, Hasenclevers, and Erskines didn't have the New Jersey iron business to themselves: "Shortly after the Revolution it was said that a man could not ride across the State in any direction without stumbling upon at least two of these old works; and in Morris County alone there were nearly a hundred iron forges in operation by the year 1777."[49] The din was terrific. Waterwheels turned the great tilt hammers used to beat impurities from the pig iron: "As they dropped, rose, and dropped again, their noise had once boomed through the quiet valleys for miles."[50]

Sometime after the Revolution, the London Company unloaded the property onto James Old. He sold it to Martin Ryerson, who for half a century, was the most important iron maker in New Jersey. His sons were less successful, however, and in 1853 they sold the estate to the Trenton Iron Company.[51] Founded in 1845 by Edward Cooper and Abram Hewitt, this had rapidly grown into the nation's largest iron works. Hewitt had often explored the New Jersey countryside searching for ores to supply their mills:

> In searching for the best ores for his Trenton mills, [Hewitt] steeped himself in the lore of the old Jersey mines, furnaces, and forges. He formed the habit of spending two or three days, when he could spare them, in excursions into the hills of the four northwestern counties, Hunterdon, Warren, Morris, and above all, Sussex. The woods—beech, oak, maple, and birch below, pine and spruce on the crests—were still deep; they were filled with the melody of waters, dashing over the gneiss and limestone rocks. In these forest recesses, following half-faded paths, he would sometimes come upon remains that seemed to speak of an older race of men. Shafts, with rusting bars and chains at the top, would suddenly open into the rocky hillsides. Heavy walls of crumbling masonry, enclosing a dilapidated waterwheel, would rise beside some brawling stream, whose waters had been diverted into a deep pit-like basin. Or some stark donjon of stone and brick, overgrown with vines and moss, with a gaping iron maw at the bottom, would close the end of a deeply rutted road, where slag lay mixed with clay and gravel—the ruins of an old furnace . . . [52]

> To Hewitt the study of these ruins was business, but he felt their romance nonetheless. They evoked the romance of woodland adventure; men had pushed into rough mountain wastes, into swamps and tangled forests, fighting the Indians and panthers as they went, not merely to take furs or clear farms, but to lay the foundations of mining and industry. Their names called up the romance of business risk; the "London Company" which built the works at Ringwood, Long Pond, and Charlotteburg in Colonial days had spent more than £54,000 before any tangible returns appeared. And there was the romance of invention. These early ironmongers had to sell their wares in the form of stoves, farm implements, kitchen utensils, hardware, and arms, and they exercised great ingenuity in devising new iron commodities . . . The iron plant at Trenton was the inheritor of all this and of the inventions made with increasing rapidity in English mills during the previous century. The ores which Cooper and Hewitt used came in the main from the Jersey hills, though some were also brought from Pennsylvania. They went to Phillipsburg, where the three tall blast furnaces—for a third was built in 1852–1853—were able to smelt at least 25,000 tons of ore annually . . .[53]

"The possibility that they might run short haunted [Hewitt's] pillow. If only he could find another property like Andover! In 1853 Ringwood, the most celebrated iron-ore property in all the East . . . came upon the market."[54] Hewitt leapt at the chance. A glance at the map showed it could easily be connected with the new Erie Railroad, and Hewitt immediately sent the company's mining expert to look at it. After receiving his enthusiastic report, Hewitt himself made a visit, pronouncing the prospects at Ringwood the best he had ever seen. Since 1763 a dozen highly productive mines had been opened there: "the Blue, the Little Blue, the London, the Cannon, the Peters (which Erskine considered the best), the St. George, and others. When Hewitt bought the place, from 300,000 to 500,000 tons of ore had been taken out, but enormous quantities were still in the ground; and he breathed easier."[55]

Suddenly the Map Looks Different

And now the map looks different; it's not the same landscape anymore. What seemed a bucolic picture of hills and lakes turns out to be an industrial site: the sound of tilt hammers booms through the quiet valleys, blast furnaces rise above the woodlands which in any case are nothing but fuel and flux, the streams are dammed for waterpower. On every side stand wire mills and rolling mills, heating furnaces and puddling furnaces, machine shops and pattern shops. Canals and railroads wire the valleys. Boats and trains carry iron ores and iron pigs. When Hewitt and his bride

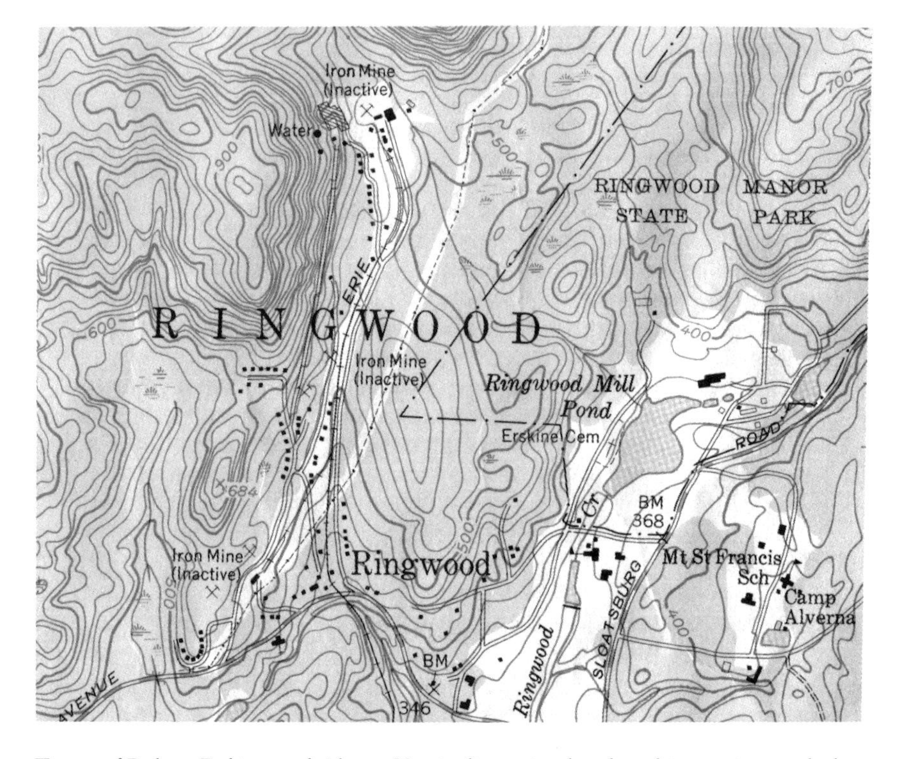

Traces of Robert Erskine and Abram Hewitt linger in abandoned iron mines and place names on this corner of the U.S.G.S.'s *Greenwood Lake Quadrangle*.

moved to Ringwood, they took over the house the old ironmonger Martin Ryerson had erected near Peter Hasenclever's mansion. They stabled their horses in what had been sorting and crushing mills. Despite the gardens for which Mrs. Hewitt became famous, there was no escaping the iron. On the terrace before the house Hewitt placed a link from the famous chain—supposedly of Ringwood metal—that the Revolutionaries had strung across the Hudson to stop the British. Beside it stood the anvil on which it had been forged, and beside that the waterwheel which had raised the anvil's great hammer.[56]

Though Hewitt was no robber baron—Henry Adams regarded him "as the most useful public man in Washington"[57]—he nevertheless held that "the consumption of iron is the social barometer by which to estimate the relative height of civilization among nations."[58] This is the spirit—with its complicated commitment to both iron *and* the public—in which, as a member of the U. S. House, he wrote the bill that in 1879 consolidated the four or five existing national surveys (King's, Hayden's,

Powell's, and Wheeler's) into the Geological Survey that would come in time to produce the *Wanaque* and *Greenwood Lake* quadrangles:

> This legislation met with much opposition. There were jealousies regarding the headship of the new Geological Survey, for Hayden had his champions as against King and Powell, who actually became the first and second heads respectively. There was the old jealousy with the War Department of civilian encroachments. There were powerful Western interests which fought to the last against any measure designed to strengthen the land laws. All Hewitt's skill and force was needed to carry the legislation in the House. He answered every objection, struck out vigorously at the "grasping corporations and overpowering capitalists" who were trying to seize our great western heritage, and showed that the Engineering Corps of the army would be kept quite busy enough in making needed military surveys. As he said, eminent scientists would not care to place themselves under young officers of the army. Above all, he explained the incalculable importance of the Western domain, the vast potentialities of wealth and growth locked up in it. In the end, he triumphantly carried the bill. Years later, when the Geological Survey hung the portraits of King and Powell in its offices, it wrote asking for that of Hewitt as well.[59]

They needn't have bothered: his portrait is inscribed in every topographic sheet the Survey produces.

What *is* it that with all its machinery of decision making (what to show, how to show it) we are laboring to see in the maps the Survey produces? It turns out to be quite straightforward. In Hewitt's words: "What is there in this richly endowed land of ours which may be dug, or gathered, or harvested, and made part of the wealth of America and of the world, and how and where does it lie?" [60] What we're laboring to see is . . . *America as a great cornucopia, a vast cupboard.*[61] Given Hewitt's liberalism this commandment has to be taken liberally—that is, with "for the benefit of all" scrawled across it[62]—but given the forces he had to contend with, it must be accepted that his liberalism is but one of the many interests inscribed on the *Wanaque* (and every other quadrangle), including those of . . . *grasping corporations* (and let's face it, the Trenton Iron Company *had* to have that ore), of . . . *overpowering capitalists* (and, however awkwardly, this *is* what Hewitt was), of . . . *the War Department* (which in fact made our map, and whose arguments for exclusive control of all surveying Hewitt was initially inclined to lean toward), of . . . *eminent scientists* (a number of whom, including Clarence King, Hewitt counted among his friends), of . . . *powerful Western interests* (which Hewitt was not entirely lacking).[63] No wonder the map does not speak in a single voice. It has never been a question of conspiracy, from the beginning we have insisted that the interest embodied in maps was neither simple nor singular.

Yet despite this polyphony, the chord that is sounded is that of a . . . *disinterested science*.[64] In its rhetorically orchestrated denial of rhetoric (the austere white margins, the tastefully subdued colors) the map seems to represent only this. Powerful Western interests, capitalism, the troubled sleep of the iron-ore poor . . . *have disappeared*. The rigorous dispassion of the survey sheet is seductive precisely in the degree to which no sign of seduction is apparent: the message of Nature Subdued (howsoever liberally the wealth is distributed)—or . . . *untouched*—is the more powerful because it seems to be spoken not by the *map* (*it* appears to say nothing, appears to *allow* the world to speak), but by Nature itself.

CHAPTER FIVE

The Interest Is Embodied in the Map in Signs and Myths

We see that this is what maps do: they mask the interests that bring them into being; this to make it the easier to accept what they say as . . . *unsaid* . . . as . . . *in the air.* This is what they do. How is it that they manage to pull this off?

Spread out on the table before us is the *Official State Highway Map of North Carolina.* It happens to be the 1978–1979 edition—not for any special reason: it just came to hand when we were casting about for an example. If you don't know this map, you can well enough imagine it, a sheet of paper—nearly 2 by 4 feet—capable of being folded into a handy pocket or glove compartment sized 4 by 7 inches. One side is taken up by an inventory of North Carolina points of interest—illustrated with photos of, among other things, a scimitar horned oryx (resident in the state zoo), a Cherokee woman making beaded jewelry, a ski lift, a sand dune (but no cities)—a ferry schedule, a message of welcome from the governor, and a motorist's prayer ("Our heavenly Father, we ask this day a particular blessing as we take the wheel of our car. . ."). On the other side, North Carolina—hemmed in by the margins of pale yellow South Carolinas and Virginias, Georgias and Tennessees, and washed by a pale blue Atlantic—is represented as a meshwork of red, black, blue, green and yellow lines on a white background, thickened at the intersections by roundels of black or blotches of pink. There is about it something of veins and arteries seen through translucent skin, and if you stare at it long enough, you can even convince yourself that blood is actually pulsing though them. Constellated about this image are, *inter alia,* larger scale representations of 10 urban places and the Blue Ridge Parkway, an index of cities and towns, a highly selective mileage chart, a few safety tips and . . . yes, a legend.

Legends

It doesn't say so, of course, but it is one all the same. What it says is: "North Carolina Official Highway Map / 1978–1979." To the left of this title is a sketch of the fluttering state flag. To the right is a sketch of a cardinal (state bird) on a branch of flowering dogwood (state flower) surmounting a buzzing honey bee arrested in midflight (state insect).

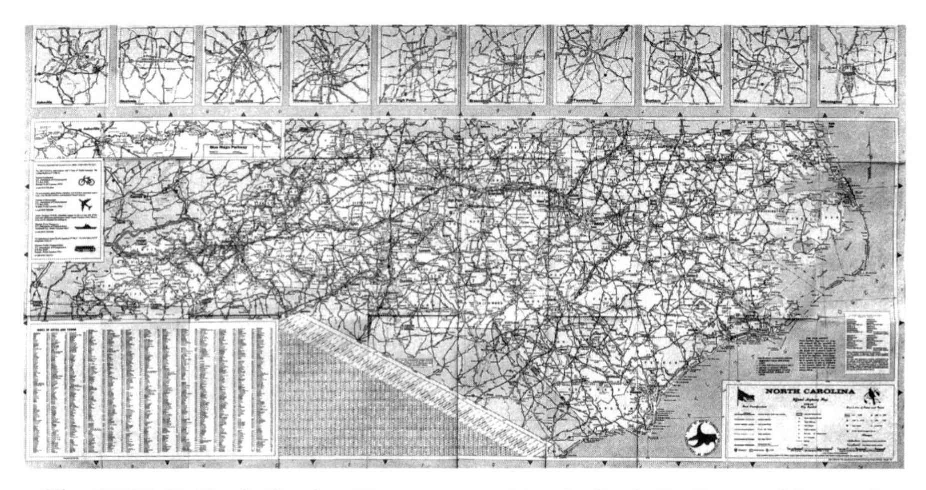

The "1978–79 North Carolina Transportation Map & Guide To Points of Interest." Unfortunately the distinctions among the pale blue, yellow, pink and white are lost in the reproduction. (North Carolina Department of Transportation)

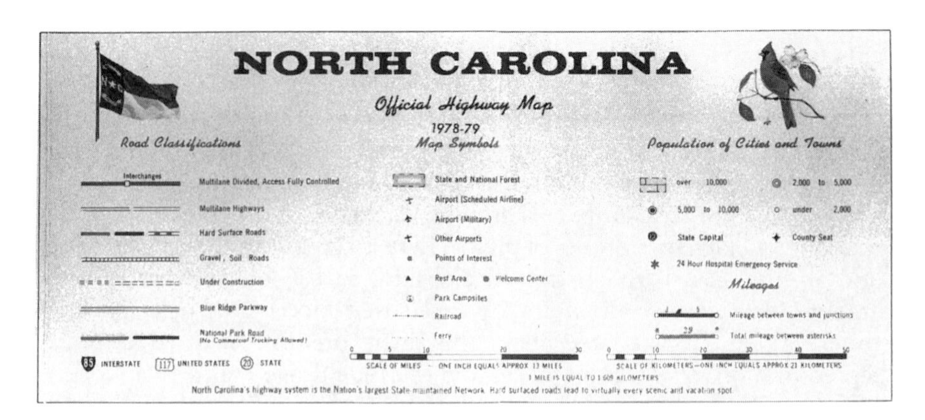

The legend block from "1978–79 North Carolina Transportation Map & Guide To Points of Interest." Again, it's too bad you can't appreciate the color. (North Carolina Department of Transportation)

Below these, four headings in red—"Road Classifications," "Map Symbols," "Populations of Cities and Towns" and "Mileages"—organize collections of marks and their verbal equivalents (thus, a red dot is followed by the words "Welcome Center"). We will return to these in a moment, but for the sake of completeness, it should be noted that below these one finds graphic and verbal scales (in miles *and* kilometers), as well as the pendent sentence, "North Carolina's highway system is the Nation's largest State-maintained Network. Hard surfaced roads lead to virtually every scenic and vacation spot."[1]

Clearly this legend—to say nothing of the rest of the map—carries a heavy burden, one that reflects aggressively the uses to which this map was put. The plural is stressed because it is a fact not so much *overlooked* (cartographers are not *that* naive), as ignored, denied, suppressed. For certainly in this case, the first and primary "user" was the State of North Carolina, which *used* the map as a promotional device (in *this* context "used" comes naturally), as an advertisement more likely than most to be closely looked at, even carefully preserved (because of its other uses), and so one given away at Welcome Centers just inside the state's borders, at Visitor Centers elsewhere, from booths at the State Fair, and in response to requests from potential tourists, immigrants and industrial location specialists. This is all perfectly obvious in "The Guide to Points of Interest" and the selection of photographs that decorate it (unless that's backwards, and the "Guide" is first of all a way of justifying the photographs, much like text in the *National Geographic Magazine*) . . . but it is no less evident in the legend itself.

Nor is it just a matter of the unavoidable presence of the state flag, flower, bird and insect—though here they are in children's encyclopedia colors—but primarily of what *else* the map's makers have chosen for the legend and the ways they have chosen to organize it (for more than one principle of order operates under even seemingly straight forward subheadings such as "Populations of Cities and Towns"). It is conventional to pretend, as Robinson *et al.* have put it, that "legends or keys are naturally indispensable to most maps, since they provide the explanations of the various symbols used,"[2] but that this is largely untrue hardly needs belaboring. Legends flare into cartographic consciousness not much earlier than thematic maps, are nonetheless still dispensed with more often than not, and never provide explanations of more than a portion of the "symbols" found on the maps to which they refer. The fact that they do not accompany topographic survey sheets (and the fact that the available legend is incomplete), or the plates of a *Rand McNally International Atlas*, makes this perfectly clear. That legends do exist for these maps—*some*place in the book, or by special order—only serves to underscore, through their entirely separate, off-somewhere-else character, exactly how dispensable they really are.

Nor is this dispensability a result of the "self-explanatory" quality of the map symbols, for, though Robinson *et al.* might insist that, "no symbol that is not self-explanatory should be used on a map unless it is explained in a legend,"[3] the fact is that NO symbol *explains* itself, stands up and says, "Hi, I'm a lock," or "We're marsh," anymore than the *words* of an essay bother to explain *themselves* to the reader. Most readers make it through most essays (and maps) because as they grew up through their common culture (and *into* their common culture), they learned the significance of most of the words (and map symbols). Those they don't recognize they puzzle out through context, or simply skip, or ask somebody to explain. A few texts come with glossaries, though like map legends these are rarely consulted and readily dispensed with. But this familiarity with signs *on the part of the reader* never becomes a property of the mark; even the most obvious, transparent sign remains opaque to those unfamiliar with the code.

It is not, then, that maps don't need to be *de*coded; but that they are by and large *en*coded in signs as readily interpreted by most map readers as the simple prose into which the marks are translated on the legends themselves. For at best legends less "explain" the marks than "put them into words," so that, should the *words* mean nothing, the legend is rendered less helpful than the map image itself where the signs at least have a context and the chance to spread themselves a little (as anyone who has "read" a map in a foreign language can attest). One way to appreciate this while approaching an understanding of the role legends actually play is to take a look at those signs on maps that don't make it onto the legend, of, for instance, this *North Carolina Official Highway Map*. Concentrating for the moment on the map image of the state proper, ignoring, that is, the little maps of the state's larger cities, the inset of the Blue Ridge Parkway, the mileage chart (the instructions for which do happen to be pasted over the map image proper, though over South Carolina, just below Kershaw), the guide to other transportation information sources, the borders and rules, and the letters, numbers and other marks that facilitate the operations of the index of cities and towns—though to pretend that any of this is half as self-evident as the signs of the map image is to miss how laboriously we have learned to interpret the architecture of this picture plane, how much we have come to take for granted—still, ignoring all this, and all the words, and somehow managing to overlook that logo of the North Carolina Department of Transportation floating on the Atlantic some twenty miles due east of Cape Fear, it is nevertheless the case that 18 signs deployed on the map image *do not appear on the legend*. That's half as many as do.

Why don't they? It's not, certainly, because they're self-explanatory. No matter how many readers are convinced that blue naturally and unambiguously asserts the presence of water, or that little pictograms of lighthouses and mountains explain themselves, signs are *not* signs for, dissolve into marks for, those who don't know the code. *Look* at these: where, in the eyes and eyebrows of Mt. Sterling, can anyone see the mountain; or, in the pair of upended nail pullers, the lighthouse at Cape Fear? Nor is there anything more "self-evident" about the use of blue for water. Not only historically has water been rendered in red, black, white, brown, pink and green,[4] but it disports in other colors on the obverse of this very map: in silver and white on the "cover" photo of Atlantic surf; in tawny–pewter in the photograph of fishing boats at anchor; in warm silver–gray in a shot of the moonlit ocean off Wrightsville Beach; and in yellow–green in the photograph of the stream below Looking Glass Falls. Only in the falls, where it indicates shadows, is there blue in any of these waters. This lack of any sort of "necessary" or "natural" coupling between blue and water proves fortuitous, for the color used to represent water on the map *image* does double-duty as background for the sheet as a *whole*, and surely we were never intended to read the circumjacent margin for a circumfluent ocean. There's no way around it: each of these *signs* is a perfectly conventional way of saying what is said ("lighthouse," "mountain," "water") —which is why the map *seems* so transparent, so easy to read. But *were* the function of the legend to explain such conventions (or at least translate them into words), then these too would belong on it, as surely as those that are there.

And if these belong there, so do the yellow tint used for "other states," the white used for "North Carolina," the thick continuous green-with-dashed-red line that asserts "National Park" and the thick continuous yellow-with-long-short-dashed-black line that stutters "county" (so long as the border isn't along or over water). These all may be equally conventional, but they are less vernacular than the blue for water and so are more likely to be misconstrued, especially on a map on which a long-short-short dashed black line mutters "state," a continuous blue line murmurs "coast" or "bank," a fine dashed red line coughs at "military reservation," a slightly thicker dashed red line says "Indian reservation" and a still thicker one proclaims "Appalachian Trail." A fine dashed line in black whispers "national wildlife refuge." A continuous line in red hints, in degrees, at the graticule.

Yet, whereas all these . . . uncommon . . . signs are absent, *on* the legend we find interpretative distinctions made among the shapes and colors of the roadsigns of the Interstate, federal and state highway systems. Does the person really exist for whom the graticule is self-evident and yet the highway signs obscure? There probably is no such human being, though doubtless there are many immured in subtleties of the highway signage system to whom the graticule and its associated cabalism of degrees and minutes is a deep mystery. What becomes gradually clear is that if the purpose of the legend ever were "explanation," everything is backwards: the things least likely to be most widely known are the very things about which the legend is reticent, whereas with respect to precisely those aspects both natives and travelers are most sure to be familiar, the legend is positively garrulous. Garrulous, but not necessarily . . . informative: the signs under the category "Road Classifications" comprise less a system than a yardsale of marks, many of which remain, despite their inclusion on the legend, "unexplained." What is one to make, for instance, of the three marks given for "Hard Surface Road"? Are we to distinguish among solid red, solid black and enclosed, dashed blue? Or are these just three arbitrary ways of designating the same reality? Suggestions of system inevitably evaporate under the heat of attention: about the time you've concluded that red is the color of federal highways, you run down US 74b in black; and by the time you've decided that unnumbered state roads are in enclosed, dashed blue, you realize you don't have the foggiest idea what that means. There are another three equally vague signs for highways under construction, and another two for multilane highways. There would seem to be an interest in portraying access (controlled or not), jurisdiction (federal or state), condition (constructed, under construction), composition (hard surface, gravel, soil) and carrying capacity (multilane or not) but not *enough* interest to force anybody to confront the graphic complexity implied by a five-dimensional code. Nor is this mess limited to the "Road Classifications" portion of the legend. Of the seven signs under "Populations of Cities and Towns," only four relate to population, and these do so without consistency. The state capital, county seats and "24-Hour Hospital Emergency Service" have individual designations confusingly related to the signs of population. Thus, the sign for "State Capital" is circular, like the signs for towns with less than 10,000 people; but the "County Seat" sign is a kind of lozenge shape. The sign for "Emergency Service" is a bright blue asterisk.

We can see your lips moving as you read this. They're saying, "What a poor excuse for a map! My five year old could do better." But that's not true. Even graduate design students collapse when confronted with a task of this complexity. The design problems alone test them (to say nothing of the . . . *cartographic* problems), but the political realities wipe them out, especially the (by now anticipated) demands of interagency collaboration (for whereas one side of our map was handled by the Department of Transportation, the other was produced by the Department of Commerce), but also the rigors of pleasing state senators *and* representatives, and the imperatives of manifesting those miniscule but vital tokens of partisanship that distinguish the map of a Republican administration from that of the Democrats. Nor is it such a poor excuse for a map. It's a fair example of the genre. It's indistinguishable, for instance, from the *Illinois Official Highway Map, 1985–1986*; from the *Michigan Great Lake State Official Transportation Map for 1974* (which makes up for the omission of its state insect by illustrating, *inter alia*, the state gem [greenstone], state fish [trout] and state stone [petoskey]); and it's a lot less weird than the *Texas—1976 Official Highway Travel Map*, which in an attempt at shaded relief manages only to look . . . badly singed. *All* the maps of the genre, and most other genres as well, are characterized by legends (like this map's) that in a more or less muddled fashion put into words map signs that are so customary as to be widely understood without the words, while leaving the map images themselves *littered* with conventions it taxes professional cartographers to put into English.

Myths

Invariably the knee-jerk reaction is either to pooh-pooh the examples, no matter how many times multiplied, as bad (as in, "Those are just *bad* maps!") or to call for a revolution in the design of their legends ("Rethinking Legends for the State Highway Map"). Both responses completely miss the point. *There is nothing wrong with the design of these legends: they are supposed to be the way they are*. This will be difficult for many to accept, but once it is understood that the role of the legend is less to elucidate the "meaning" of this or that map element than to function as a sign in its own right, this conclusion is even more difficult to . . . *evade*. Just as the bright blue asterisk signifies "24-Hour Hospital Emergency Service," so the legend as a whole is itself a signifier. As such, the legend refers not to the map (or at least not directly to the map), but back, through a judicious selection of map elements, to that to which the map image itself refers . . . *to the state*. *It is North Carolina that is signified in the legend, not the elements of the map image*, though it *is* the selection of map elements and their disposition within the legend box that encourages the transformation of the legend into a sign. It is a sign only a cartographer (or graphic designer) could fail to understand. Others receive in a glance, naively or otherwise, this sign of North Carolina's subtly mingled . . . *automotive sophistication*, *urbanity* and *leisure opportunity*. Apprehended this way, the legend makes sense. The headings in red—heretofore so bizarre—appear now as *headlines* to a jingoist text. Under the fluttering flag, appear the words, "Road Classifications." *Plural*. North Carolina's road system is so rich, that one classification can't handle it. And across the legend, under the bucolic branch *cum* bird (read "rural," read "traditional values") and the bee if you can see it (read "hard working" [read "no unions"]), the words, "Populations of Cities and Towns." Cities *and* towns . . . *and* birds and bees.[5] It is almost too much, though as it says on the 1986–1987 edition of this map, "North Carolina has it all."

It certainly has a lot of whatever it is. Look at those road signs! Their proliferation can no longer be seen as a manifestation of graphic and taxonomic chaos, though, but as a sign insisting that roads really *are* what North Carolina's all about. The sign's abundant density supports the presumption of the headline and justifies the proximity of the flag. That there are more signifiers than signifieds is no longer a mystery to be explained, but part of the answer to the question, "Does North Carolina *really* have a lot of roads?" It's the graphic analogue to the assertion in black at the bottom of the legend box that reads: "North Carolina's highway system is the Nation's largest State-maintained Network."[6] What the roads connect, of course, are all those cities. It's wonderful the way it takes seven signs and four lines to unfold the complexities of what the cartographer can't help observing is but a four-tier urban hierarchy. Again, it's the graphic equivalent of a remark from the governor's letter on the other side of the map about "booming" cities. Hey: this is a *hip* state (though bucolic), urban, urbane, sophisticated (but built on traditional values). The whiff of sophistication is heightened by the kilometer scale, so *European*, almost risqué (though it's carefully isolated in the lower right hand corner of the legend under the heading, "Mileages"). Roads and cities: roads *to* and *from* cities, that is, exactly the desideratum for someone looking to locate, say, a plant somewhere in the South. Modern, in other words, up-to-date. But as the bird and branch and honey bee remind us . . . *not off the wall.*

And yet it's not all work either. In between, in between moments, in between the roads and the cities and towns, in between the *signs* for the roads and the cities and towns, under the innocuous heading "Map Symbols" (which from its central position also casts its net over all the map signs on the legend), may be found the signs for fun, *clean* fun, *good* clean fun, but still fun: "Park Campsites," "State and National Forest," "Welcome Center," "Rest Area" and "Points of Interest," to say nothing of the signs for still other ways of getting around, ferries, railroads and *three* kinds of airports. Led by that bright green forest sign that visually lies at the center of the legend (read "parks"), this heterogeneity speaks of caring for people ("*Welcome* Center," "*Rest* Area") and is the graphic version of the remainder of that black sentence that sums up the legend (and is counterpoised at the bottom against "North Carolina" at the top): "Hard surfaced roads [for which there are three signs] lead to virtually every scenic and vacation spot."

Wow! It's almost overdone. Had it been done up slick by some heavy duty design firm, it would have been overdone. But here, it's just hokey enough to seem sincere. *It is sincere.* We don't believe for a minute anyone sat down and cynically worked this out, carefully offsetting the presumptuousness of the overheated highway symbolism with the self-effacing quality of the children's encyclopedia colors. But this is not to say that with this legend we are not in the presence of what Barthes has called "myth"—a kind of "speech" better defined by its intention than its literal sense.[7] Barthean myth is invariably constructed from signs that have been already constructed out of a previous alliance of a signifier and a signified. An example, an especially innocuous one, is given by the reading of a Latin sentence, "*quia ego nominor leo,*" in a Latin grammar:

> There is something ambiguous about this statement: On the one hand, the words in it do have a simple meaning: *because my name is lion.* And on the other, the sentence is evidently there in order to signify something else to me. Inasmuch as it is addressed to me, a pupil in the second form, it tells me clearly: I am a grammatical example meant to illustrate the rule about the agreement of the predicate. I am even forced to realize that the sentence in no way *signifies* its meaning to me, that it tries very little to tell me something about the lion and what sort of name he has; its true and fundamental signification is to impose itself on me as the presence of a certain agreement of the predicate. I conclude that I am faced with a particular, greater, semiological system, since it is co-extensive with the language; there is, indeed, a signifier, but this signifier is itself formed by a sum of signs, it is in itself a first semiological system (*my name is lion*). Thereafter, the formal pattern is correctly unfolded: there is a signified (*I am a grammatical example*) and there is a global signification, which is none other than the correlation of the signifier and the signified; for neither the naming of the lion nor the grammatical example is given separately.[8]

The parallels with our legend are pronounced. On the one hand, it too is loaded with simple meanings: *where on the map you find a red square, on the ground you will find a point of interest.* But as we have seen, the legend little commits itself to the unfurling of these meanings, even compared to the map image on which each is actually named—"Singletary Lake Group Camp" or "World Golf Hall of Fame." The appearance of the red square on the legend thus adds nothing to our ability to understand the map. Instead it imposes itself on us as an assertion that *North Carolina has points of interest;* in fact, it speaks *through* the map *about* the state. Yet, as in Barthes' example, this assertion about North Carolina is constructed out of, stacked on top of, the simpler significance of the red square on the legend, namely, to be identified with the words, "Points of Interest."

We thus have a two-tiered semiological system in which the simpler is appropriated by the more complex. Barthes has represented this relationship this way.[9] In our case, at the level of language we have as signifier the various marks that appear on the legend: the red square, the black dashed line, the bright blue asterisk. As signified we have the

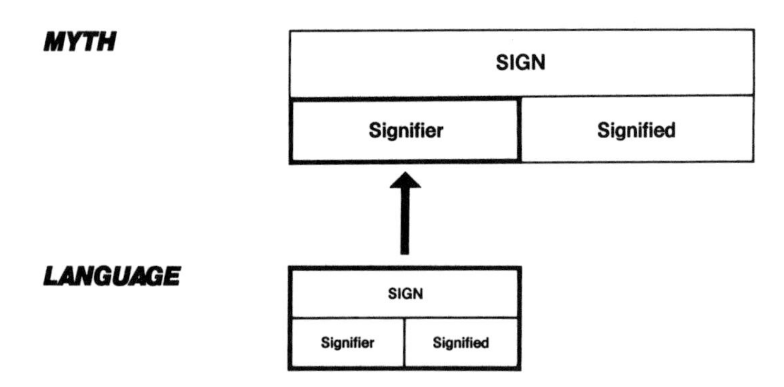

Signified and signifier are conjoined in the sign, the whole of which is seized by myth to be the signifier in its second-order seismological system. Barthes cautions that the spatialization here of the pattern of myth is only a metaphor. (Redrawn from the diagram, p. 115, of Roland Barthes, *Mythologies*, Hill and Wang, New York, 1972.)

respective phrases: "Points of Interest," "Ferry" and "24-Hour Hospital Emergency Service." Taken together, the marks and phrases are *signs*, things which *in their sign function* are no longer usefully taken for themselves (there is no red square 350 yards on a side at Singletary Lake) but as indicative of or as pointing toward something else (a point of interest called Singletary Lake Group Camp). Collectively, these signs comprise the legend, *but this in turn is a signifier in another semiological system cantilevered out from the first*. At this level of myth we have as signified some version of what it might mean to be in North Carolina, some idea of its attractiveness (at least to a specifiable consumer), a concept signed also in the photos decorating the other side of the map, in the governor's message, in the "Motorist's Prayer," a concept we could call . . . *North Carolinaness*. The signifier is of course the legend appropriated from the level of language by this myth to be its sign. Insidiously, this myth is not required to declare itself in language. This is its power. At the moment of reception, it evaporates. The legend is only a legend after all. One sees only its neutrality, its innocence. *What else could it be? It is after all a highway map!*

Indeed. *And so it is*. It is precisely this ambiguity that enables myth to work without being seen (that enables the Van Sant map, the *Wanaque Quadrangle*, and this highway map to mask the interests that brought them into being). Perched on top of a primary semiological system, myth resists transformation into symbols (which makes it hard to put into words, hence . . . *hard to talk about*). As a legend or a map or a photograph, it retains always the fullness, the presence, of the primary semiological system to which it is endlessly capable of retreating. What viewed obliquely appears as an advertising slogan, confronted directly is the blandest of legends, so that the slogan, still ringing in one's ears, is apprehended as no more than the *natural* echo of the facts of the map. It is in this way that *North Carolinaness* comes to be accepted as *an attribute of the terrain* instead of being seen as the promotional posture of state government it actually is. This constitutes, in Barthes' phrase, "the naturalization of the cultural":

> This is why myth is experienced as innocent speech: not because its intentions are hidden—if they were hidden they could not be efficacious—but because they are naturalized. In fact, what allows the reader to consume myth innocently is that he does not see it as a semiological system but as an inductive one. Where there is only an equivalence, he sees a kind of causal process: the signifier and the signified have, in his eyes, a natural relationship. This confusion can be expressed otherwise: any semiological system is a system of values; now the myth consumer takes the signification for a system of facts: myth is read as a factual system, whereas it is but a semiological system.[10]

Not seen as a semiological system: this is the heart of the matter. Of all the systems so not seen, is there one more invisible than the cartographic? The most fundamental cartographic claim . . . *is to be a system of facts*, and its history has most often been written as the story of its ability to present those facts with ever increasing accuracy. That, as we have seen, this system can be corrupted everyone acknowledges: none are more vehement in their exposure of the "propaganda map" than cartographers. But as we have also seen, having denounced this usage they feel but the freer in passing off their own products as anything other than the semiological systems they have no choice but to be. It may no longer appear that an official state highway map is quite such a system of facts as it might have seemed; but this is essentially a consequence of our presentation. Outside of this context, a highway map is accepted as inevitable, as about as natural a thing as can be imagined. Its presence in glove compartments, gas station racks (even if today they must be paid for) and the backs of kitchen drawers is . . . *taken for granted.* Yet as we have shown, even so innocent a part of the map . . . *as the legend* . . . carries an exhausting burden of myth, to say nothing of the *prayer*, *governor's message*, *photographs* and *other paraphernalia* cosseting the map image proper.

Nor does the map proper—if we can refer to such a thing—escape the grasp of myth. On the contrary, it is more mythic precisely to the degree that it succeeds in persuading us that it is a natural consequence of perceiving the world. A state highway map, for instance, is unavoidably . . . *a map of the state:* that is, an instrument of state polity, an assertion of sovereignty. There was, for example, no need from the perspective of the driver to have colored yellow the states contiguous to North Carolina on its highway map. There was no *real* need to have shown the border. It is not, after all, as though the laws regulating traffic changed much at its borders, though to the extent they do, the map is silent.[11] At this level of language, the map, like the legend, *seems* to proffer vital information; but it's an impression hard to sustain—*there is too little information to make what's provided useful*. Like the legend, the map in this regard makes no sense. From the perspective of myth, however, this delineation of the state's borders is of the essence. Though many will see in this only the most dispassionate neutrality (what could be more natural than the inclusion of the state's borders on its highway map?), there is nothing innocent about the map's affirmation of North Carolina's dominion over the land in white. Not only has effective territorial control long been dependent on effective mapping, but it is among other things the repetitive impact of the image of the territory mapped that lends credence to the claims of control (and hence the extensive logogrammatic application of the state's outline to seals, badges and emblems). Who would question the pretensions, the right to existence, the *reality* of North Carolina? Look! There it is on the map! The 1.6 *million* copies of the 1986–1987 edition of this map constitute 1.6 million assertions of the state's sovereignty, assertions which, however, at the moment of being noticed have the ability to fade back into the map where their appearance is taken entirely for granted, overlooked because expected . . . *naturally* . . . part of the surface.

Which is myth's way: the map is always there to deny that the significations piled on top of it are there at all. It is only a map after all, and the pretense is that it is innocent, a servant of that eye that sees things as they really are. But outside the world of speech, outside the world of maps, states carry on a precarious existence: little of nature, they are much of maps, for to map a state is to assert its territorial expression, to leave it off is to deny its existence. Only when it is admitted that a state unrecognized (unmapped) is scarcely a state, that it is the determination (choice) of people to acknowledge (map) it that endows with substance an assertion of statehood, or not to acknowledge (map) it that relieves it of significance, is it possible to comprehend the anger directed at maps that acknowledge the independent existence of Bophuthatswana, Transkei, Ciskei and Venda; that deny the independent existence of Taiwan; or that, for that matter, run county borders through Indian reservations, such as those of Swain and Jackson counties through the Cherokee Indian Qualla Boundary on the North Carolina highway map.[12] It is not that the map is right or wrong (it is not a question of accuracy), *but that it takes a stand while pretending to be neutral on an issue over which people are divided*.[13] Nor is it that those angered have confused the map with the terrain, but that they recognize what cartographers are at such pains to deny, that, like it or not, willingly or unwillingly, because *au fond* maps constitute a semiological system (that is, a system of values), they are ever vulnerable to seizure or invasion by myth. They are consequently, in all ways *less like the windows through which we view the world and more like those windows of appearance from which pontiffs and potentates demonstrate their suzerainty,*—not because cartographers necessarily want it this way, but because, given the manner in which systems of signs operate, they have no choice.

Paradoxically, it is an absence of choice founded on choice alone, for to choose is to reveal a value, and a map is a consequence of choices among choices. That the choice of mapping Bophuthatswana as an independent nation reveals a political attitude is something many will readily concede. But *all* choices are to a degree political, and it is no less revealing to choose to map *highways*, for this also is a value. That it would be difficult to produce a state highway map without highways is admitted, but there is no injunction on the state to map its roads anymore than there is for it to map the locations of deaths attributable to motor vehicles, or the density of cancer-linked emissions from internal combustion engines, or the extent of noise pollution associated with automotive traffic.[14] It would be satisfying to live in a state that produced 1.6 million copies of such maps and distributed them free of cost to travelers, tourists, immigrants and industrial location specialists, but states find it more expedient to publish maps of highways. North Carolina *does* publish the *North Carolina Public Transportation Guide*—a highway map-like document displaying intercity bus, train and ferry routes—but it printed 15,000 copies of the most recent edition, less than a hundredth as many maps as it printed of its highways.[15] Not an advertisement, the public transportation map was produced without the assistance of the Department of Commerce. Could this be why, unlike the highway map among whose blond hikers, swimmers, golfers and white-water enthusiasts no blacks appear, blacks figure so prominently on the public transportation map? Here blacks buy intercity bus tickets, get on city buses, and in wheel chairs get assisted into specially equipped vans. The reek of special assistance is like sweat: "Many of you have requested information on how to make your trip without using a private automobile. Because of these requests. . . ." But there is nothing of this tone on the highway map. There was never any *need* to have requested a highway map: it, after all, is . . . *a natural function of the state*. Everything conspires to this end of naturalizing the highway map (even the map of public transportation), of making the decision to produce such a map seem less a decision and more a gesture of instinct, of making its cultural, its historical, its political imperatives transparent: you see through them, and there is only the map, innocent, of nature, of the world as she really is.

Codes

It is, of course, an illusion: *there is nothing natural about a map*. It is a cultural artifact, a cumulation of choices made among choices every one of which reveals a value: not the world, but a slice of a piece of the world; not nature but a slant on it; not innocent, but loaded with intentions and purposes; not directly, but through a glass; not straight, but mediated by words and other signs; not, in a word, as it is, but in . . . *code*. And of course it's in code: *all* meaning, *all* significance derives from codes, *all* intelligibility depends on them. For those who first encountered their codes in the breakfast cereal box—little cardboard wheels arbitrarily linking letters and numbers—this generalization of the idea may occasion some disquiet. It shouldn't. When you wear a tie to work, you're dressing in code. When you frown, you're expressing in code. When you open a door for a lady—or wait for a man to open a door for you—you're gallanting in code. When you type or scribble, you're writing in code. Human languages are probably the most elaborate and complex codes we're familiar with—and the dictionary just a big clumsy breakfast cereal toy—but there are sublinguistic codes of incredible sophistication (those danced by Ginger Rogers and Fred Astaire) and supralinguistic codes of deep subtlety (such as the conventions underwriting the structure of James Joyce's *Ulysses*). Usually a number of different codes are used simultaneously (this is a text). Fred and Ginger were placed in settings, dressed, wore their hair a certain way, gestured, spoke and sang as well as danced and all this was coded.[16] The code of conventions structuring *Ulysses* cannot be encountered outside the code of English in which it is embedded. There is even a code of codes: mime, for example, is forbidden the code of words, and in general the arts are distinguished by a code whose elements are other codes. It has long been a hallmark of cartography that it speaks in art as well as science.

More technically a code can be said to be an assignment scheme (or rule) coupling or apportioning items or elements from a conveyed system (the signified) to a conveying system (the signifier). The highway code is paradigmatic of the way this works. On the one side are intentions (she intends to turn), promises (Holly Springs will be encountered 3 miles down this road) and commands (not to pass, to stop, to go). On the other side are gestures (a hand stuck straight out the driver's window), words and numbers ("Holly Springs/3 miles"), and lights and lines (a red traffic light, a solid yellow line down the middle of the road). The intentions, promises and commands are elements of the system conveyed: *signifieds* (content). The gestures, words, numbers, lines and lights are elements of the system conveying: *signifiers* (expression). The code (the rule—in this case, the Law) *assigns* the latter to the former, couples them. In so doing, it creates a *sign*.

An important distinction is being made here. The sign is *not* in the gestures or the lights, the words or the numbers: it is *not* the signifier. Nor is the sign in the intentions, promises or commands: it is *not* the signified. The sign exists solely, utterly and exclusively in its correlation (established by the code, the rule, by custom, by the law). There is nothing, for instance, inevitable (necessary) in the relationship between a driver sticking his arm straight out the left window and his intention to turn left (and in fact it has been largely supplanted by the flashing of lights on the left side of the car), any more than there is between a driver pointing to heaven and his intention to turn right (though doubtless there was some historical contingency that made it customary). These might, however, quite readily change places (may have already in some parts of the world), so that a left arm stuck straight out a left window signalled an intention to turn right and one stuck straight up signalled an intention to turn left: it would make no difference from the perspective of communication, for the meaning is in the code, and the new code could be as readily mastered as the old. Signs, in other words, are the creatures of codes with the loss of which they are rendered—like fat—into their constituent components, disembodied signifieds separated from insignificant signifiers. It is the codification in which the sign adheres, nothing else. Or, as Umberto Eco puts it:

> A sign is always an element of an *expression plane* conventionally correlated to one (or several) elements of a *content plane*. Every time there is a correlation of this kind, recognized by a human society, there is a sign. Only in this sense is it possible to accept Saussure's definition according to which a sign is the correspondence between a signifier and a signified. This assumption entails some consequences: **a** *a sign is not a physical entity*, the physical entity being at most the concrete occurrence of the expressive pertinent element; **b** *a sign is not a fixed semiotic entity* but rather the meeting ground for independent elements (coming from two different systems of two different planes and meeting on the basis of a coding correlation).[17]

Because signs neither have physical existence (unlike the signifier) nor permanence, they are frequently referred to as *sign-functions*, or in Eco's words:

> Properly speaking there are not signs, but only *sign-functions* . . . A sign function is realized when two *functives* (expression and content) enter into a mutual correlation; the same functive can also enter into another correlation, thus becoming a different functive and therefore giving rise to a new sign-function. Thus signs are the provisional result of coding rules which establish *transitory* correlations of elements, each of these elements being entitled to enter—under given coded circumstances—into another correlation and thus form a new sign.[18]

This is not a game of words. Nor is the vocabulary important. What *is* important is the notion that signs, or sign-functions, or symbols—what they are called *does not matter*—are realized *only* when coding rules bring into correlation two elements or items (or functives) from two domains or systems (the one signifying, of expression; the other signified, of content) and that *whenever* there is such a correlation, there is a sign. You may call this resulting sign an icon. You may call it a pictogram. You may call it a word. You may call it an index. You may call it a symbol. You may call it a piece of sculpture. You may call it a sentence. You may call it a map. You may call it New York City. In every case, whatever else it is, it is, *in its sign function*, also a sign, that is, a creature of a code.

No signs without codes. This must be insisted upon: that is, there are no self-explanatory signs; no signs that so resemble their referents as to self-evidently refer to them. They are inevitably arbitrary, inevitably reveal . . . *a value*. Jonathan Culler says:

> Saussure, taking the linguistic sign as the norm, argues that all signs are arbitrary, involving a purely conventional association of conventionally delimited signifiers and signifieds; and he extends this principle to domains such as etiquette, arguing that however natural or motivated signs may seem to those who use them, they are always determined by social rule, semiotic convention. Peirce, on the contrary, begins with a distinction between arbitrary signs, which he calls "symbols," and two sorts of motivated signs, "indices" and "icons," but in his work on the latter he reaches a conclusion similar to Saussure's. Whether we are dealing with maps, paintings, or diagrams, "every material image is largely conventional in its mode of representation." We can only claim that a map actually resembles what it represents if we take for granted and pass over in silence numerous complicated conventions. Icons seem to be based on natural resemblance, but in fact they are determined by semiotic convention.[19]

Once the superordinate role of the convention (the rule, the code) is accepted, it becomes easy to explain how what "self-evidently" resembles a river on a map equally "self-evidently" resembles veins on a diagram of the circulatory system, without invoking complicated principles of metaphor (not that these might not have been operant in the genesis of the sign). It is not that the reader thinks, "Oh, yes, the deoxygenated blood is relatively bluer than that in the arteries, *and* under a clear blue sky the surface of rivers often seems blue; *and* both veins and arteries carry (whatever "carry" means) liquids in a branching (see "tree") network (see "net," see "weaving"), sooo, let's see, that means. . ." This is not how it happens at all. What happens is that the reader finds himself or herself in an entirely distinct coded circumstance *all at once*. At the level of language, the diagram of the circulatory system is decoded without reference to the codes of the map, and *vice versa*. There is certainly no question of *resemblance* with respect to which Barthes notes, that it would be in any case a resemblance *to an identity* (the *identity* of the river, the *identity* of the vein), an identity "imprecise, even imaginary, to the point where I can continue to speak of 'likeness' without ever having seen the model,"[20] as those do who justify this sign for veins because "they look like veins" without ever having seen a vein (without having seen a hepatic vein, without having seen an inferior vena cava), or the sign for a river (the Colorado) because "it looks like a river" (the Thames? the Cuyahoga?) without having seen it (without having seen where the Colorado trickles all but dry into the Gulf of California). It is not a matter of resemblance: the blue line is a blue line. It is the code that does the work, not the signifier. If there is involved an iconicism, it is always at the level of the structure of the system (it is analogic not metaphoric). It is less the *blueness* of deoxygenation that says "veins" than the *simultaneous* redness of the arteries, their *characteristic* jointure at the extremities, and their *perfect parallelism*; it is less the blue-between-black lines that says "river" than its *characteristic* form, its *characteristic relationship* to other forms (other rivers, mountains, roads, towns and oceans); so that "veins" can as easily be read in black or gray, and "rivers" in diagrams of drainage basins and maps of flood insurance purchase. To say that it is the code that does the work, not the signifier, is just another way of saying that it is the code that makes the sign, not the mark.

Ten Cartographic Codes

So it is the *codes* upon which one must fasten if the map is to be *de*coded (or if a map is to be *en*coded). It is possible to distinguish at least 10 of these (doubtless there are others), which the map either exploits, or by virtue of which the map is exploited. Neither class is independent of the other, and no map fails to be inscribed in (at least) these 10 codes. Those that the map exploits are termed *codes of intrasignification*. They operate, so to speak, within the map: at the level of language. Those by virtue of which the map is exploited we term *codes of extrasignification*. These operate, so to speak, outside the map . . . *at the level of myth*.

Among the codes of intrasignification, five at least are inescapable, the *iconic*, the *linguistic*, the *tectonic*, the *temporal* and the *presentational*. Under the heading *iconic* we subsume the code of "things" ("events"), with whose relative location the map is enrapt: the streets of Genoa, rates of death by cancer, exports of French wine, the losses suffered in Napoleon's Russian campaign, airways, subways, the buildings of Manhattan, levels of air pollutants over six counties in Southern California, the rivers, roads, counties, airports, cities and towns of North

Carolina. The iconic is the code of the inventory, of the world's fragmentation: into urban hierarchies, into hypsometric layers, into wet and dry. The *linguistic* is the code of the names: the *Via Corsica*, the *Corso Aurelio Saffi*; trachea, bronchus and lung cancer, white males, age-adjusted rate by county, 1950–1969; *France, Amérique du Nord; Moscou, Polotzk;* DME chan 82 St John VSJ 113-5; Cortland St World Tr Ctr N RR Path; the Graybar Building, the Seagram; Orange County, Reactive Hydrocarbons; Cape Fear River, US 421; Pasquotank, Cherry Pt., Winston-Salem, Hickory. The linguistic is the code of classification, or ownership: identifying, naming, assigning. The relationship of these things in space is given in the *tectonic* codes: in the *scalar*—in the number of miles (or feet) encoded in every inch—and in the *topological*—in the planimetry of cities, the stereometry of mountain ranges, the projective geometry of continents, the topographometry of the field traverse, the simple topology of the sketch map giving directions to the cocktail party. The tectonic is the code of finding, it is the code of getting there: it is the code of getting. Because there is no connection, no communication, except in time, the codes of filiation are *temporal*, codes of duration, codes of tense. The *durative* establishes the scale, the map's *durée* its "thickness": as the map of rates of death from cancer, *1950–1969*, is "thicker" than the *1978–1979* North Carolina highway map, which is "thicker" than the map of reactive hydrocarbons, *6 a.m. to 9 a.m., July 22, 1979*. The durative reveals (or hides or is mute about) lapses in cosynchronicity. The *tense* says . . . *when:* some maps are in the past tense ("The World of Alexander the Great"), others in the future tense ("Tomorrow's Highways"), but most maps exist in the present ("State of the World Today"), or, if they can possibly get away with it, the aorist: no duration at all (no thickness), out of chronology (not lost—just out of it), free of time (such maps attain to myth at the very level of language).

Each of these codes—iconic, linguistic, tectonic and temporal—is embodied in signs with all the physicality of the concrete instantiation of the expressive pertinent element. On the page, on the sheet of paper, on the illuminated display with its flashing lights, these concrete instantiations are ordered, arranged, organized by the *presentational* code: they are . . . *presented*. Title, legend box, map image, text, illustrations, inset map images, scale, instructions, charts, apologies, diagrams, photos, explanations, arrows, decorations, color scheme, type faces are all chosen, layered, structured to achieve speech: coherent, articulate discourse. It is a question of the architecture of the picture plane, what's in the center and what's at the edge, what's in fluorescent pink and what's in the blue of Williamsburg, whether the paper crackles with (apparent) age or sluffs off repeated foldings like a rubber sheet, whether the map image predominates or the text takes over. It is never, even at the lowest level, a question merely of escaping the stigmas of paranomia and aphrasia, dysphemia and idiolalia, dyslogia and cacology. At the very bottom it is a question of fluency and eloquence, and soon enough of vigor and force of expression, of rhetoric, of polemic, for wherever it may begin the code of presentation soon enough carries the map *out* of the domain of intrasignification into that of extrasignification, into that of the society that nurtures it, that consumes it . . . *that brings it into being*.

Among the codes of extrasignification five again are inescapable, the *thematic*, the *topic*, the *historical*, the *rhetorical*, and the *utilitarian*. All operate at the level of myth, all make off with the map for their own purposes (as they made the map), all distort its meaning (its meaning at the level of language) and subvert it to their own. If the presentational code permits the map to achieve a level of discourse, the *thematic* code establishes its domain. *On what shall the map discourse? What shall it argue?* Though it is precisely the thematic code that has dictated their appearance on the map, from the perspective of the reader, the theme is experienced as a latency inherent in the "things" *iconically* encoded *in* the map: roads, for instance, it is a map of roads and highways; it asserts the significance of roads and highways (if only by picturing them, if only by foregrounding them); its theme is Automobility (the legitimacy of Automobility). Or it is a general reference map, a map of hydrography and relief carved into political units and plastered with railroads and towns, that is, a map, of a landscape smothered by humanity, tamed subdued (the red railroads—sometimes black—inevitably reminiscent of the bonds by means of which the Lilliputians restrained Gulliver), its theme is Nature Subdued. And precisely as the thematic code runs off with the icons, so the *topic* code (with a long *o* from *topos*, place, as in *topography*, not *topicality*) runs off with the space established by the tectonic code, turns it from space to place, gives the map its *subject*, bounds it (binds it), names it (via the linguistic code), sets it off from other space, asserts its existence: *this place is*. Just so the *historical* code. Only it works on the time established in the map by the temporal code. Are there bounding dates to the map's *durée*? Then the historical code appropriates them to an era, assigns it a name, incorporates it in a vision of history (it establishes the map's subject . . . in time). So an archeological map of Central America acquires the title, "Before 1500/Pre-Columbian Glory," one of 19th century plantation crops, political units, selected urban places, cart roads, railroads and battles the title, "1821–1900/Time of Independence," yet another of similar subjects (though with the addition of a sign for refugee centers) the caption "1945–Present/Upheaval and Uncertainty."[21] There is no time that cannot be reduced to these sequacious causal schemata, absorbed into these . . . platitudes, made comfortable and safe because grasped, understood.

If the thematic code sets the subject for the discourse, if the topic and historical codes secure the place and time, it is the *rhetorical* code that sets the tone that, having consumed the presentational code, most completely orients the map in its culture (in its set of values), pointing in the very act of pointing somewhere else (to the globe) to itself, to its. . . *author*, to the society that produced it, to the place and time and omphalos of that society—the more dramatically as the aspect of the globe toward which it points is alien, is exotic, *i.e.*, can have its title set in a typeface that mimics . . . *bamboo*. It is a code of jingoisms, a code that beats its chest like Tarzan, a code of the sort of subtle chauvinisms that encourages the *National Geographic* to call it a "road" on its map of the Central Plains, 1803–1845, but to call it a "*cart* road" on its map of Central America, 1821–1900.[22] But after all, it is an "American" map, that is, a map that reflects the genius of the *North* Americans, or at least those north of the Rio Grande (for according to the *National Geographic* the ancient Maya had but "*trade* routes" and even the Camino Real was just a "trail"); and, if only because it *is* the mapping society, the mapping society stands at stage center, with all the others in the wings. For the rhetorical code, the mere existence of the map is a sign of its higher culture, its sophistication:

A television weatherman points to a map. At the same time, it points back to him, establishing and emphasizing his modernity, sophistication, and thus his reliability. In turn, this flatters our sense of self-esteem for having selected this station over others. This map is all but consumed by its rhetorical functions.

the map is rhetorical *au fond*, and for this reason no map can eschew it. It is like clothing: even not to wear it is to be caught in the net of meanings woven by the code of fashion. To attempt to shed the rhetorical code is but to shout the more stridently through it: it is its very disregard for the subtler aspects of the code of presentation that so completely characterizes the publisher of *The Nuclear War Atlas* as "socially conscious"[23]; it is nothing other than their violations of "good taste" that allows us to read the editors of *The State of the World Atlas* as angry.[24] Their *subversion* of the power of the rhetorical code amounts to a bold proclamation of their rhetorical stance (cartographic nudism, cartographic streaking, cartographic punk), the very opposite of the position occupied by the United States Geological Survey, which, as we have seen, obscures its stance beneath a rhetorically orchestrated *denial* of rhetoric (dressing itself in the style of science). Elsewhere the map will dress in the style of Art. Or in the style of the Advertisement. Or in the Vernacular (the North Carolina Highway map). The rhetorical code appropriates to its map the style most advantageous to the myth it intends to propagate. None is untouchable. All have been exploited.

As the map itself is finally exploited, picked up bodily by the *utilitarian* code to be carted off for any purpose myth might serve. A professor of curriculum and instruction, commenting on the availability of state highway maps for secondary classroom use, remarks, "It has the governor's picture on it. You can get as many as you want." It is here that the academic model of the map with its scanning eyes and graduated circle-comparing minds breaks down most completely. It has no room for the real uses of most maps, which are to possess and to claim, to legitimate and to name. What great king, what emperor, what great republic has failed to signal its coming of age by the mapping of its domains? Whatever the pragmatic considerations (they are, after all, maps that speak also at the level of language), it has inevitably also been an act of conspicuous consumption, a sign of contemporaneity as well as wealth and power, a symbolic manifestation of the rights of possession. *These* are the uses of maps as certainly as it is the most important function of maps in geographic journals to certify the geographic legitimacy of the articles they decorate.

Despite our slight foray, the anthropology of cartography remains an urgent project: what *are* all those maps actually used for? Signs, badges, tokens, emblems, billboards, gestures, leases, deeds, wallpaper, pretty picture. Do not say "not *this* one"—not *that* topographic survey sheet—for as surely as you do, it will turn out to be the one with the most heinous agenda, it will be the one lying about the Love Canal, the one suppressing the missile silos.[25] Whatever else this might be, it is not a gesture of disinterested curiosity . . . it is one of exploitation. But, as we have seen, what else to make of Survey sheets? Dressed in their button-down white shirts and suitable ties, these, in their metered regularity (so many sheets per unit area), their sensible no-nonsense layout, their methodical tiling, their obsessive coverage, know no other code. "To catalogue," Barthes notes, "is not merely to ascertain, as it appears at first glance, but also to appropriate."[26] In the end, survey sheets differ little enough from maps of . . . *military targets*.

Intrasignification

The map, then, is comprehended in two ways. As a medium of *language* (in the broadest sense) it serves as a visual analogue of phenomena, attributes, and spatial relations: a model on which we may act, in lieu or anticipation of experience, to compare or contrast, measure or appraise, analyze or predict. It seems to inform, with unimpeachable dispassion, of the objects and events of the world. As *myth*, however, it refers to itself and to its makers, and to a world seen quite subjectively through their eyes. It trades in values and ambitions; it is politicized. Signing functions that serve the former set of purposes we have termed *intrasignificant;* those

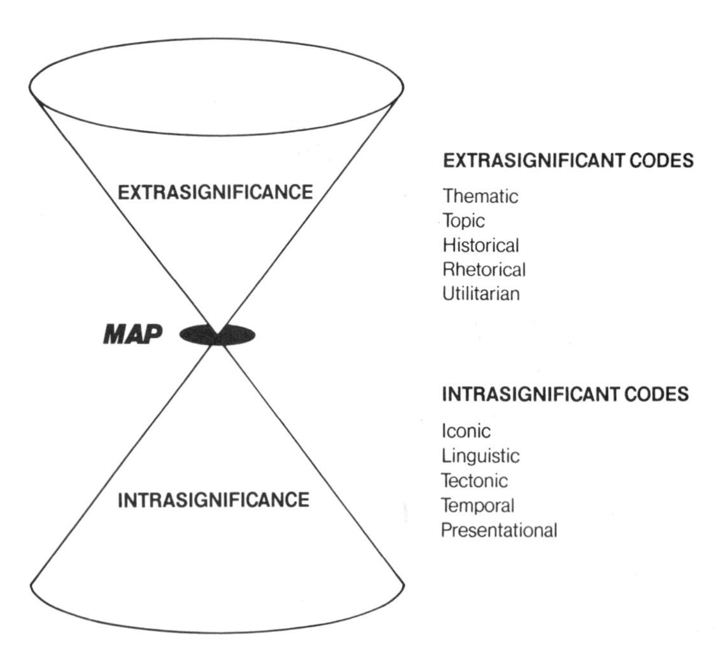

The map as a focusing device between the domains of extra- and intrasignification: the map gathers up the constituent signs governed by the codes of intrasignification so that they will be able to act as signifiers in the sign-functions governed by the codes of extrasignification—which specified them in the first place.

which serve the latter, *extrasignificant*. Whereas intrasignification consists of an array of sign functions indigenous to the map and which, taken jointly, constitute the map . . . *as sign*, extrasignification appropriates the complete map and deploys it . . . *as expression* in a broader semiotic context. The map acts as a focusing device between these two planes of signification, gathering up its internal or constituent signs and offering them up collectively . . . *as a map*. But what effers from the map is not substantially different from what is afferent upon it—these have simply been repositioned in the semiological function—and, whereas extrasignification exploits the map in its entirety, *we have seen how the initiatives of myth extend to even the most fundamental and apparently sovereign aspects of intrasignification, and are ultimately rooted in them*. How, then, does this happen?

The map is the product of a spectrum of codes that materialize its visual representations, orient these in space and in time, and bind them together in some acceptable form. The actions of these codes are, if not entirely independent, reasonably distinct. *Iconic codes* govern the manner in which graphic expressions correspond with geographic items, concrete or abstract, and their attendant attributes. A *linguistic code* (occasionally two or several) is extended to the map to regulate the equivalence of typographic expressions, and via the norms of written language, a universe of terminology and nomenclature. As the space of the map is configured by *tectonic codes*—transformational procedures prescribing its topological and scalar relations to the space of the globe—*temporal codes* configure the time of the map in relation to the stream of events and observations from which it derives. The diversity of expressions that constitute the map are organized and orchestrated through a *presentational code* that fuses them into a coherent cartographic discourse. Here we turn to each of these in turn.

Iconic Codes

Iconicity is the indispensable quality of the map. It is the source and principle of the map's analogy to objects, places, relations, and events. In its capacity as geographic icon, the map subsumes a remarkable variety of visual representations and the codes, both general and specific, that underwrite them; yet the degree of iconicity evident in the map as a whole is not uniformly echoed among its constituents. The dot that represents a town is not iconic in the same way as the intricately shaped area representing a city; the blue line representing a river is not iconic in the same sense as the blue line representing a county road or, for that matter, a shoreline. Pursued far enough, every icon is seen as the product of two procedures: a *symbolic* (substitutive) *operation* that provides the basis of its representative potential, and a *scheme of arrangement* that yields its specific and individual form. The balance struck between these has frequently been the canon by which we judge representations as symbolic (of the town, for example) or iconic (of the city); and although this distinction will not be abandoned here, it will be applied with extreme care. No symbol is *totally* arbitrary unless it can be stripped entirely of connotation (an unlikely and undesirable prospect), and no icon is motivated free of convention because *representation without convention is not possible*. We can only say that some representations are more explicitly iconic or symbolic in function; that media of cultural exchange—maps in particular—serve as proving grounds where iconic representations gradually acquire symbolic status through a process of reiteration and cultural distension.

The iconicity of Hermann Bollmann's *New York Picture Map* is so powerful that its representational conventions virtually disappear from view.[27] On inspection, the picture plane . . . *melts away*, and our attention falls into a landscape of tangible urban forms: streets, sidewalks, roofs, facades, doors, windows. It seems so literal, so transparent to interpretation, so . . . *natural* that it is difficult to accept as a highly conventionalized and essentially symbolic representation. Yet without our conventions of pictorial rendering, this arresting image would be opaque and meaningless.[28] Make no mistake: iconicity, as Bhattacharya has explained, is the product of a spatial transcription[29]; and its derived form is an arrangement of marks in relationship to one another and to the space they occupy. The icon is motivated not by a monolithic precedent form but by the formal and necessarily *spatial arrangement* it would transcribe on the page, and it can only materialize through a *transcriptive* procedure. This procedure, in Bollmann's map, turns out to be extraordinarily elaborate: involving 67,000 photographs taken with specially designed cameras, an axonometric projection spread in two dimensions by a calculated widening of streets, and, according to the map's jacket, "several unique devices which remain his secret." It emerges from a tradition of representation that is distinctly Western and intensively codified, and it speaks through a familiar (to us) regime of symbolic principles: lines demark intersections of planes and boundaries between solid and void; certain organizations of lines denote rectilinear volumes; recurring tonal patterns denote illuminated forms.

Thus, to describe iconicity as a simple matter of visual likeness (as if this *could* be a simple matter), or as a formal correspondence between expression and referent, is to mystify its explanation and divorce it entirely from cultural enterprise. Iconicity derives from our ability to transcribe arrangements in space and mark them out in conventional symbols—in other words . . . *to map them*. This ability is as fully realized in a drawing by da Vinci as in a Swiss topographic map, where the natural

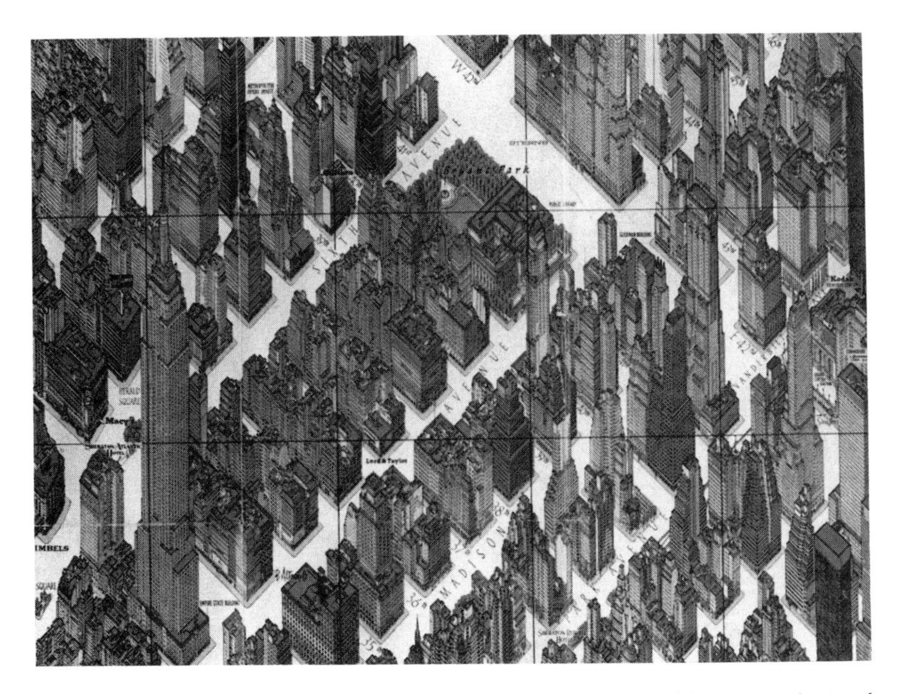

A portion of Bollman's Manhattan. This compelling icon is an elaborate synthesis of Western representational conventions.

landscape—like Bollmann's urban landscape—is portrayed as a complex and continuous icon, bathed in light and rendered with the consummate authority of an iconism as richly meaningful for its audience as for its maker.

A map of population distribution produced by the U. S. Bureau of the Census has some of this same pretense.[30] Substitute night for day, luminosity for reflectivity, and city form for architectural or geomorphic form, and we have an equally credible—if more remotely viewed—icon of human settlement. But the symbolism of this map is more explicit, and less uniform; in fact it embraces several distinctly different representative principles. Urbanized areas, like Bollmann's office towers and Imhof's mountains, enter the map as geographic icons, shaped by the space of the features themselves transcribed onto the graphic plane. Isolated cities and towns, however, enter as geometrically pure squares and circles regardless of their geographic shape; they have undergone an abstraction conventionalizing their form and enacting their status as symbols.[31] Beyond and between these, symbols are disengaged from exact spatial correspondence and are referred to features that are in themselves abstractions. In the first instance, form is given as the consequence of the

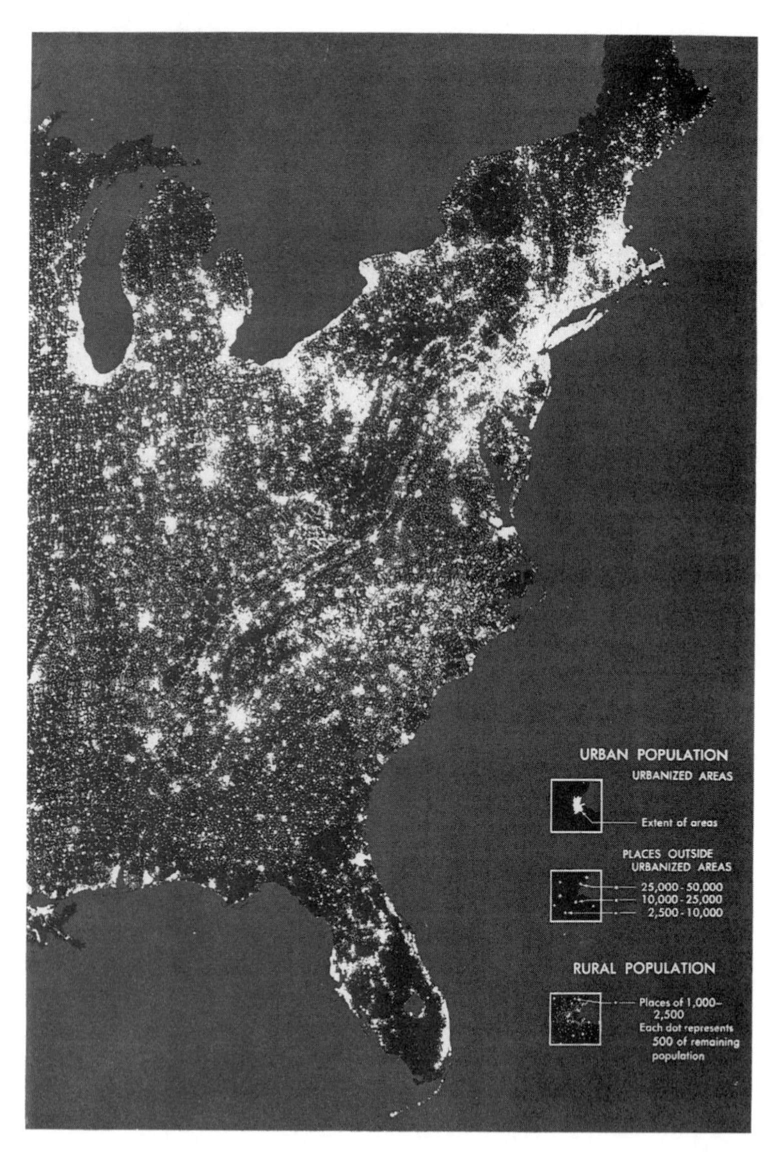

From a lexicon of graphic symbols, a geographic icon. While significant in itself, each mark, like a point of color in a Seurat painting, is subservient to the impression of the whole. (From *Maps for Americans, by Morris M. Thompson, published by the U.S. Department of the Interior, 1979.)*

feature's spatial extension and the topological transformation that implants it on the page. Symbolism remains characteristic: white is city, dark blue is water (or foreign terrain), black is neither. In the second instance a formal symbolism is activated: white *square* is city or white *circle* is city. In the third instance, symbols are fixed not only in form but in value as well, and they acquire a limited but necessary mobility within a scheme that treats them not as localized occurrences (in which case they have no literal meaning) but as elements of a comprehensive system to be interpreted *en masse*. This map is truly a *tour de force*, an exemplar of cartographic representation deploying an arsenal of significant strategies from the most abstract and conventionalized to the most geographically constrained and overtly iconic. Although we might expect, from this description, a baffling and practically indecipherable stew of signs, what we have instead is a remarkably legible and coherent representation, one that correlates strongly with a photographic representation of the same phenomena.[32] Profoundly different principles of symbolism merge, almost seamlessly, in an icon that eschews the formal consequences of their application and takes their distribution as the basis of its own.

Signs formed, rather than just characterized, independently of geographic space are free to engage in formal metaphor. A lighthouse is signed with an ornamented triangle or an outlined circle and a complement of rays, a mine with an occluded dot or an emblematically crossed pick and shovel. Extracted from map context, these signs are icons in their own right—but icons of what? The triangular lighthouse sign and the circular mine sign are ostensible abstractions of their phenomenal counterparts and, regardless of their degree of abstraction, they remain icons insofar as they maintain a structural correspondence with them. But the circle and rays sign is iconic only in respect to the light, not the lighthouse, and it represents by virtue of a part-for-whole substitution. The pick and shovel sign (with no regard for technological currency) represents min*ing* rather than mine by substituting artifact for process. These last two examples are conventional metaphors, parallels to which abound in maps.[33] They differ from the icons of urban form and symbols of city size in not referring literally to the phenomena they represent. They anticipate interpretation by singling out connotations and presenting them as surrogate icons. Icon is proffered, and taken, as symbol.

In signs which *are* geographically conformal, metaphor operates through *characteristic*. Green symbolizes trees and blue, water in our maps with the same conviction they did in the childhood drawings that implanted these metaphors in our vocabulary, never mind drought, autumn, and acid rain, and never mind the cubic miles of eroded silt that choke our rivers. In the map, our forests glow with the robust verdure of a perpetual spring afternoon, and even the Mississippi shines with a pristine Caribbean blue. These metaphors proclaim the map as *ideal* (or at least hyperbole), at once an analogue of our environment and an avenue for cultural fantasy about it. False coloration is hardly restricted to remotely-sensed imagery; it is characteristic of *all* our maps, which it dresses in . . . *the most reassuring tones*.

The iconic code of the map is a complex mix of more specific codes—potentially any established or even *ad hoc* code of graphic representation, provided it either is or can be conventionalized. The map seems to have assimilated the entire history of visual communication, maintaining an immense pool of representational techniques and methodologies from which it draws freely, with little preference or prejudice, and which it augments through continual invention and recombination. Although this inventory is far too extensive to be catalogued here, we can summarize the object of its application. The map is an icon, a visual analogue of a geographic landscape. It is the product of a number of deliberate, repetitive, symbolic gestures, carefully arranged and explicitly or implicitly referred to elements of a content taxonomy. Formal items—the discrete elements of iconic coding—may be shaped within the space of the map, in which case their symbolism and metaphorical potentials are characteristic, or preformed and imposed on the map, activating formal symbolism and formal metaphor as well. The diversity of cartographic expression far surpasses that of written language or any other medium of practical exchange; but map signs are only as diverse as our abilities to interpret them, and their formation is as firmly prescribed by the confines of our own visual culture, the array of conventions that dictate how we may equate marks and meanings. The iconic code of the map is the sum of its various conventions of graphic representation; the comprehensive icon of the . . . *map image* . . . is the synthesis of their actions.

Linguistic Codes

It is difficult to imagine a map without language. However separate the evolution of iconic and linguistic representation, the map has, for millennia, embraced both. External to the map image, language assumes its familiar textual forms: identifying, explaining, elaborating, crediting, cautioning. Its main role, though, lies within the map image and in its interpretive template, the map legend. Like graphic marks, typographic marks sign the content of the map on different yet complementary grounds.

In the legend, semantic connections are made between classes of graphic images or image attributes and linguistic representations of the phenomena to which they refer. In this capacity, the legend acts as interpreter between the unique semiological system of the individual map and the culturally universal system of language so that on seeing a red circle, for example, we may hear the words "Welcome Center" (even if we're not entirely sure what that means). In translating graphic expression to linguistic expression we make the map literate and its meanings subject to literary representation and manipulation. It seems our compulsion and need to do so.

Within the map image, linguistic signs address not only what things are called ("Lake") but also what they are *named* ("Superior"). Thus identification is a matter of both *designation* and *nomenclature*. Much of our geographic nomenclature carries a residuum of designation, as in "Union *City*," "Youngs*town*," "Louis*ville*," "Pitts*burgh*"; but it is practically obligatory with respect to natural features. One word, "river" for instance, may occur hundreds of times within a single map image. The cartographer who would erase this redundancy, however, finds that rivers are no longer distinguishable from creeks, nor lakes from reservoirs. Here language is not just naming features, but *illuminating content distinctions* that have, for whatever reason, escaped iconic coding.

If the function of language in maps were simply toponymic, we could assume that the linguistic signifiers themselves, if recognizably formed and correctly arranged, would be fixed in meaning. This is clearly not the case. Within the map image, elements of visible language serve as counterparts to iconic signs, overlapping their content and spatial domains and echoing their iconic properties. In the map image, entire words and arrangements of words are given iconic license, generating a field of linguistic signs best likened to concrete poetry. Letters expand in size, increase in weight, or assume *majuscale* form to denote higher degrees of importance. Stylistic, geometric and chromatic variations signal broad semantic divisions. Textual syntax is largely abandoned as words are stretched and contorted and word groups rearranged to fit the space of their iconic equivalents. Clearly this code invokes more than the disposition of phonetic archetypes.[34]

It's not that the map rejects the ground rules of textualized language; if it did, it would quickly degenerate to a vehicle for newspeak or nonsense. Even seemingly absurd statements like "Lac Champlain Lake" and "Rio Grande River" are grammatically functional in a bilingual or multilingual culture. *What this code gains in the cartographic context is nearly unrestricted access to the means of iconic coding.* Among attempts to produce maps entirely from linguistic signs, the more successful have been cognizant of these means[35]; and in even the most familiar maps the field of typographic signs, taken on its own, visualizes the geographic landscape in much the same way as the field of graphic signs. The map is simultaneously . . . language *and* image. As word lends icon access to the semantic field of its culture, icon invites word to realize its expressive potentials in the visual field. The result is the dual signification virtually synonymous with maps as well as the complementary exchange of meaning that it engenders. The map image provides a context in which the semantics of the linguistic code are extended to embrace a variety of latent iconic potentials[36]; to the same end, it imposes a secondary syntax that shapes entire linguistic signifiers into local icons.

Tectonic Codes

To reiterate: a code is an interpretive framework, a set of conventions or rules, which permits the equivalence of expression (a graphic or typographic mark) and content (forest, population of less than 1,000 persons, or multilane limited-access highway). In effect, a code *legislates* how something may be construed as signifying, as *re*presenting, something else. In this respect signs are encoded in formation and decoded in interpretation; and it is only through the mediation of a code that signification is possible.

Each map employs a tectonic code—we have discussed this—a code of construction, which configures graphic space in a particular relation to geodesic space.[37] This code effects a *topological* transformation from spheroid to plane in sign production and plane to spheroid in interpretation. It has a *scalar* function as well, logically separable from the topological but not practically independent of it. Whereas the role of this code as representative principle is evident, its content and expression are less so, *because both of these functives are abstract space.* The tectonic code governs a *sign function* that has as its content a *topology* and as the product of its action a *correlative topology*. If cartographic projections and scales have not been widely recognized as codes, it is not, as we have seen, because, they are difficult to formulate as such (since in most cases they can be reduced to concise mathematical expressions, they are indeed *more easily* formulated than the iconic and linguistic codes). Rather it is because they do not in themselves produce material imagery: they offer space for space, abstraction for abstraction, *and their work is not visible until it is subjected to iconic coding.* The mesh of graticule lines cradling the map image is not the tectonic code itself, but an *icon* of the topology acted upon by this code. Nor is it obligatory to render this topology: frequently it is manifest only in the shape and disposition of features, and, when it is visualized, it serves primarily as a referencing system to implement the literalization or numeralization of space.

Yet as we have seen, this code traffics in spatial *meanings*, and the messages it allows us to extract from the map are messages of distance, direction, and extent. It shapes and scales the graphic plane in such a way that these messages emerge from the map image. While iconic and linguistic codes access the semantic field of geographic knowledge, *the tectonic code provides their syntactical superstructure;* this is the code through which we signify not what, but *where*. In molding the map image, the tectonic code allows it to refer to the space that we occupy and experience; and inevitably it is laden with our . . . preconceptions about that space. It cannot therefore surprise to find the map projection at the center of political controversy, pretending as it does to validate our cultural centrism and objectify our territorial aims. It has these potentials because it allows us to view the world as we choose—as much or as little of it as we like, from whatever vantage point we like, and with whatever distortions we like—and, even though we know better, it nevertheless projects an aura of ubiquity and authenticity. It can do so because we recognize it as the only thing exact—if in the most limited sense—in a practice that propagandizes exactitude as if this were the reason for its existence.

Temporal Codes

"Every map is out-of-date before it's printed." This adage is a staple of the cartographic office. It is customarily dragged out for the benefit of the novice, held up as a fact of life (like death or taxes), and then put aside as an inevitable consequence of the complexities—of the *paradox*—of the mapping process. If meant seriously, it's as a barb at the sluggishness of the mapping bureaucracy—every member of the bureaucracy except, of course, the cartographer. But for the most part it evokes laughter or sentient smiles rather than angst (*let's not get too wound up over it; we said out-of-date, not obsolete*), and it's really not the sort of thing that cartographers lose sleep over. (It just makes them . . . uneasy.)

Somehow we've gotten the idea that maps have nothing to do with time. We'll indicate a date of publication, and perhaps a time frame for data collection, but that's about as far as it goes—and these gestures have more to do with the status of the map as a document than with any issue of *map time*. We shrug that off, if a bit nervously, because we've learned to make maps in the terms they can resolve: *anything that changes fast enough to render the map genuinely obsolete before it can reach its audience doesn't belong in the map in the first place*. The map is opaque to these things: it filters them . . . out. That's partly a function of scale: maps are macroscalar and macroscopic, and, after all, we *are* mapping mountains and not the pebbles inching down their slopes. But the things we're increasingly interested in mapping don't have this short-term permanence at any scale; they're more in the nature of *behaviors* than geographic fixtures.[38] These interests may inspire new map forms, but they haven't forced us yet to admit that maps *embody* time as surely as—in fact *because*—they embody space. It remains conventional to think of the map as either a snapshot—in time but not of it; something with time evaporated out of it (as the Van Sant)—or as akin to a 3-hour exposure of Grand Central Station in which actions, events and processes disappear, and all that register are *objects of permanence* (as implied by the durative code of the Geological Survey). We *may* be aware of emplacing time in the photograph, and even of permanence as the arbitrary consequence of this act, but we refuse to extend these understandings to the map. Time remains a . . . *hidden dimension*, a cartographic *Twilight Zone*. But the map *does* encode time, and *to the same degree* that it encodes space; and it invokes a temporal code that empowers it to signify in the temporal dimension. That the action of this code on temporal attributes should be explained by the action of two subcodes, which parallel those acting on spatial attributes, is hardly surprising. The map employs a code of *tense*, concerning its temporal *topology*, and a code of *duration*, which concerns its temporal *scale*.

Tense is the direction in which the map points, the direction of its reference in time. It refers to past, to present (or a past so immediate as to be taken as present), or future—relative, of course, to its own temporal position. So we have maps in the past tense (*East Asia at the time of the Ch'ing Dynasty*), maps in the present tense (the *1986–1987 North Carolina Transportation Map*), and maps in the future tense (of tomorrow's weather, or a simulation of nuclear winter). We also have temporal *postures*, the fantastic map (of Middle Earth, Dune, or Slobbovia) with its present and past separate, but not entirely detached, from our own; and the allegorical map (*The Map of Matrimony, The Gospel Temperance Railroad Map, The Road to Hell*[39]) that proclaims itself *atemporal* or *eternal* and, thus, presumes the *aorist* of the Greek. As maps slide into the past they become *past* maps ("antique" is a term reserved for past maps of some virtue or special appeal) where they continue to refer to *their* pasts, presents and imagined futures. The posture of the facsimile and the counterfeit is one of position rather than reference, the facsimile admitting (if only in a whisper) of its true temporal position.

The distinction between present and past is always difficult. A map positioned in the last century is obviously *past*—or is it? The physiographic map of 1886 is past by virtue of its cultural references—its references to the state of physiographic knowledge or the state of graphic representation in 1886—not by virtue of its content, which we still insist we can scale into . . . immutability. Erwin Raisz's physiographic maps, interleaved among the pages of the modern atlas, appear transported there from another time—*and they are*—but we take them all the same as maps *of the present*.[40] Without a more stable yardstick, the passage of cartographic time is marked off in editions. For the atlas these are accelerated by the pace of political and developmental change and braked by the constraints of map production; for the topographic map it's modulated by the intensity of localized activity; and with the digital database it's fixed in a perpetual, virtual present.[41] Meanwhile, as we have seen, the Survey quadrangle expresses time—that between the map in hand and its predecessor—with a violent purple tint that says . . . *these things are new.* Cherished globes have been sacrificed to garage sales and flea markets, the megabuck atlas is becoming an art investment, and we even have a class of disposable maps (with a lifespan roughly equal to that of a newspaper) characterized not so much by their funk as their anticipated, and almost immediate, obsolescence. We are increasingly conscious of the distance between present tense and past tense; and while it's still remarkably elastic, it is—as everyone tells us—shrinking fast.

The *durative* code of the map operates on the scalar aspect of time. As spatial scale constitutes a relationship between the space of the map and the space of the world, temporal scale constitutes a relationship between the time of the map and the time of the world; that is, the map embraces this or that span of world time, it has a certain thinness, or thickness. For example, an electronic map of traffic density in downtown Raleigh. In 1 minute, it plays out on a color graphics terminal the events of an entire day. This map has a *temporal scale* that is the ratio of one interval (a minute) to another (24 hours), or 1:1440. *It's just like a spatial scale.*[42] Of course, that was a convenient example. Consider instead a newcomer to Raleigh mapping out his environment from a bus window. It's Saturday afternoon, and he's just boarded the South Saunders bus at the central transfer point on Martin Street.

4:51 It will be 4 minutes before the bus leaves. Outside a few dozen people sit around on benches talking, reading newspapers, or just waiting, enjoying the Spring sun slanting between the banks and commercial buildings lining the Fayetteville Street Mall. In one direction the Mall slides down to the glassed and steel-trussed Convention Center. At the other end, three blocks away, the turquoise dome of the State Capital bulges over its massive oaks. The view in both directions is fragmented by the Mall's decor: saplings, floral planters, a scattering of sculptures, a clock mounted on a mirrored kiosk. There are seven other passengers on the bus now, one of them thrusting his hand relentlessly into a box of candied popcorn. The next seat bears five knife slits, and here and there a *nom de plume* stands out in the faded graffiti: "Catbird," "The Non Stop Crew," "Woogie Tee."

4:55 The bus rolls from the curb, stops abruptly as another nudges in front of it, then groans away. The street is compressed by gray and beige walls rising a half dozen stories from the sidewalk. At eye level the bus reflects dimly in the plate glass of old shop fronts. Everything is in shadow.

4:57 A right turn onto Blount Street. To the left, aging warehouses catch the sunlight head on. One of them announces its renovation. The next block's been leveled on both sides, and, to the right, a sea of asphalt and windshields foregrounds the city's nucleus of office towers. Several blocks of shotgun shacks, verandas crowded with laundry lines and painted metal chairs, then the expanse of South Street slashed clear around Memorial Auditorium, an imposing chunk of institutionalized Art Deco.

4:59 The bus dips beneath the Shaw University pedestrian bridge, careens right onto Smithfield, and stops beside a tiny parkette of juniper. Here Wilmington and Salisbury streets merge into Highway 50 and zip off in six grass-trimmed lanes of new pavement toward the Garner suburbs. As cars burst past in both directions, the driver weighs his odds . . .

5:01 Past the commuter's raceway, the bus rattles over a set of railway tracks and the backside of Memorial Auditorium jumps across the right windows. Swinging left onto old Fayetteville Street, it stops below a cascade of terraces capped by an archetypal red brick elementary school. Directly across the street, a project sprawls out, sheathed with brown wood siding and decorated in spray-bomb cursive. One person leaves the bus, and two teenage girls hoist a stroller through the front doors.

5:03 To the right a fresh canopy of leaves spreads over the weathered monuments of Mount Hope Cemetery, and to the left the project gives way to squared-off little homes. The bus wheels right onto Maywood and the small homes persevere, gradually brightening. On the neighborhood basketball court, a girl in a pink jumpsuit buries a fifteen-footer.

5:06 The bus lurches across a graded swath of red soil that imprints the future widening of South Saunders Street and brakes to a halt opposite Earp's Seafood. It turns right onto South Saunders, then left at Carroll's Used Tires, then right again onto Fuller. A stretch of tidy compact houses ends suddenly at Lake Wheeler Road. A tire swing (one of Carroll's?) hangs outside the near window. Several passengers disembark here; one boards and is recognized. "How you doing?" "All right!"

5:08 The bus cuts right onto Lake Wheeler Road and descends a long grade. To the left a high chain link fence tracks its descent, staking out the boundary of Dorothea Dix Hospital. To the right a precipitous slope tumbles into a clutter of rooftops and ahead Raleigh's best downtown panorama spreads over the windshield. At the foot of the grade, the road dovetails back into South

Saunders where a column of plaster hens files across the eaves of R. B.'s Chicken 'n' Ribs.

5:11 Passing the entrance to the Dorothea Dix grounds, the bus stops in front of Heritage Park (another housing project but far more ambitious than the one on Fayetteville Street). Three riders step out cradling their afternoon purchases, and a right turn onto South Street aims just off the downtown core. Another descent, bottoming out below a closely set pair of railway trestles, then a quick rise and a confusion of lanes. With Memorial Auditorium a block ahead the bus pivots left onto McDowell.

5:13 On the left, a parking lot, then a Chevy dealership. On the right, another parking lot, then another, then another. Cars everywhere. No people, just cars, waiting. The downtown towers against the right window and then disappears behind a four story parking deck. A cluster of satellite dishes crowds together on an office rooftop.

5:15 At the corner of McDowell and Martin the green expanse of Nash Square spreads out over the driver's left shoulder. A handful of people wander, without apparent intention, across the park. Turning right, the bus squeezes between the walls of Martin Street, gets lucky at the Salisbury traffic light, and then slips against the curb. The doors open. It's still 79° outside, but in the shadows it feels cooler.

If the bus hadn't returned to Martin Street, there would be nothing especially spatial about this experience; it unfolds *in time* as a sequence of impressions, and its spatial quality remains latent until it reconnects with its point of origin and becomes a *closed traverse*, At that point everything witnessed becomes . . . synchronous and the previously confounded immigrant exclaims, "I know where I am!" (implying that, to some degree, "I know where I've been"). Space has been surrounded and captured (unlike the tenuously connected scenes lingering along its perimeter, beyond the grasp of its closure): *time has collapsed into space*. It is still present in the map, but . . . *as space*.[43] In Minard's *Carte Figurative* of Napoleon's Russian campaign,[44] time is literally distance, marked out by the rhythm of falling boots and shrinking roll calls. Less dramatically, but more explicitly, the "Driving Distance Chart" at the back of the AAA road atlas recognizes each segment as simultaneously a spatial interval (255 miles) *and* a temporal interval (5 hours and 20 minutes).[45] Curiously—or perhaps predictably—it also tries to subvert its identity as a map, even proclaiming itself a "chart" (read, "*not* a map"), but it still looks like a map and it still functions as one.

We can pretend that the dimensions of the map are entirely synchronic, that it has no diachronic quality except as a specimen of technical or methodological evolution; but every cartographer who has

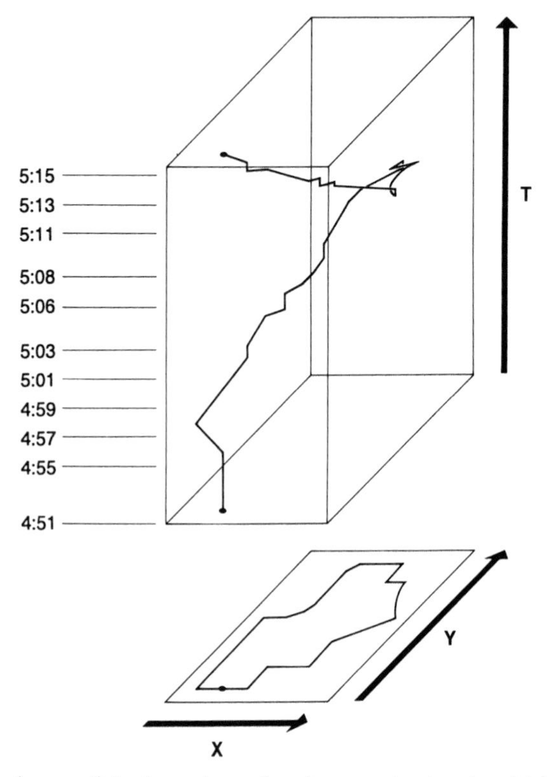

A spatio-temporal map of the bus trip, and a planar projection in which the temporal dimension has been collapsed to zero thickness. Space emerges as the product of synchronization (temporal flattening) and the closure of movement.

grafted a new road onto an old, or dropped the still warm symbols of his latest research onto the cool plate of a 20-year old base map, knows better. The potential for anachronism is vast; and sometimes it runs amok, as in the maps that drag our earliest continental explorers across a fabric of 48 American states or 10 Canadian provinces (*Native states? What native states?!*). Time is always present in the map because . . . it is inseparable from space. Time and space are alternative and complementary distillations, projections of a space/time of a higher dimensional order. We cannot have a map without thickness in time unless we can have a map without extension in space. We cannot squeeze time out of the map, only onto it.

Presentational Codes

The time of the map, the space of the map, the phenomena materialized in this framework, and the roster of terms and toponyms cast into it are

. . . not the map. Expressed through a complex of iconic and linguistic marking schemes, they become the content of the *map image;* but the *map,* as we have already pointed out, is much more than this solitary image orphaned on its audience's doorstep. The map image is accompanied by a crowd of signs: titles, dates, legends, keys, scale statements, graphs, diagrams, tables, pictures, photographs, more map images, emblems, texts, references, footnotes, potentially any device of visual expression. The map gathers up this *potpourri* of signs and makes of it a coherent and purposeful . . . *proposition*. How these signs come together is the province of a presentational code, which takes *as content* the relationship among messages resident in the map and offers *as expression* a structured, ordered, articulated and affective display: a legitimate discourse.

The more apparent aspects of this code are intrasignificant. It acts on the structure of the map, dividing and proportioning the space of the page, staking out the prospective geometry of blocks, columns, channels and margins. It proceeds from the primacy of the rectangle, echoing our Euclidean systemization of environment (objects, rooms, buildings, streets, cities), use (trims, folds, stacks, racks, packages, pigeonholes) and reading itself. Within this latent superstructure the ingredients of the map are laid out, ordered by a positional scheme fixing relations of sign to sign and sign to ground and imposing on the map a *program*, a discursive strategy. Discourse is articulated through emphasis (large or small, prominent or subdued) and elaboration (the relative complexity of signs, the intricacy of their meaning).

But the presentational code works beyond schemes of graphic organization. As it acts on the map as a whole, its effects are *manifest in the whole map;* and some of these are aimed clearly toward extrasignification. The map has a discursive tone: soft/loud, even/dynamic, complacent/agitated, polite/aggressive, soothing/abrasive. The majority of "good" maps position themselves on the left side of these oppositions, more conscious of the demands of . . . professional decorum than sensitive to those of their subject matter—or perhaps their intent is to pacify by shading even the most urgent and disturbing themes into Muzak (the reverse is equally incongruous: some of the most thematically mundane maps bludgeon their viewers with symbols that weigh on the page like musket balls). The map also reflects on itself. It asserts its status among maps in its consumption of resources as mean or lavish, frugal or conspicuous: the scale of its effort, the virtuosity of its craft, its opulence of color, material sensuality, the abundance of surface left unprinted, its sheer size. These gestures are all the more obvious in the atlas, where they can pile up into an object of palpable thickness and weight. So at one extreme we have the Park Avenue hedonism of the *World Geo-Graphic Atlas*, bound by a cloth-wrapped and gold-imprinted cover a quarter of an inch thick and framed by striking end papers that sprawl over nearly five square feet.[46] At the other extreme we have the grim imperative of *The Nuclear War Atlas:* an anti-atlas in the form of a Marxist tabloid, a document one could well imagine run off after hours on a hand-cranked press and thrust at nervous yuppies on street corners, or nailed to a senator's door.[47] Government maps are especially status-conscious, announcing the cost of their printing or the percentage of recycled pulp in their stock in an effort to disarm the bellicose taxpayer. The map also proclaims its alignment: its professional camp (a Cartographer's map as opposed to a Designer's map), its institutional allegiance (a National Geographic map as opposed to a Bartholomew, a Rand McNally as opposed to an AAA) and occasionally the method and aesthetic of its author (a Bollmann map of Manhattan as opposed to an Anderson). It has a projective aspect as well: it's prepared for a particular audience. It is manufactured for the urbane or the profane, the casual or the attentive, for those at ease with maps or for the cartophobic, for the executive or the mercenary, the well-to-do or the student, the sighted or the blind. It speaks in *their* language: in clinical ascetic, in hot-color High-tech, in journalistic cartoon, in Country and Western, or suburban rec-room.

The presentational code of the map can't be explained as a simple set of rules for graphic organization, especially without defining *whose* rules. Its action is not limited to the structural aspects of presentation or confined to affairs of visual priority and reading sequence (not at least until computers produce maps *for* computers). The map isn't a debating club exercise; it's set firmly in the real world, where the abstraction of structure, order, and articulation cannot be cut away from issues of aesthetics or even belief—any more than the grammar of this text can be separated from its meaning or the attitudes and values of its author.

Sign Functions

Maps are about relationships. In even the least ambitious maps, simple presences are absorbed in multilayered relationships integrating and disintegrating sign functions, packaging and repackaging meanings. The map is a highly complex supersign,[48] a sign composed of lesser signs, or, more accurately, a synthesis of signs; and these are supersigns in their own right, systems of signs of more specific or individual function. It's not that the map conveys meanings so much as *unfolds* them through *a cycle of interpretation* in which it is continually torn down and rebuilt; and, to be truthful, this is not really the map's work but that of its user, who creates a wealth of meaning by selecting and subdividing, combining and recombining its terms in an effort to comprehend and understand. But however elaborate, this is not an unbounded process. Inevitably, it has a lower bound, the most particular sign function that resists decomposition into constituent signs, and an upper bound, the integral supersign of the entire map that accesses the realm of extrasignification; and between these extremes it is stratified. Twofold stratifications have been repeatedly proposed,[49] and widely accepted, but these don't go far enough. If we intend to explain how the map generates and structures the signing processes by virtue of which it is a map, then we need as least four strata or levels of signification: the *elemental*, the *systemic*, the *synthetic*, and the *presentational*.

At the *elemental* level, visual occurrences (marks) are linked with geographic occurrences (features) in the set of germinal sign functions announced, if incompletely, by the map legend. At the *systemic* level, signs (supersigns) are composed of similar elements, forming systems of features and corresponding systems of marks. At the *synthetic* level (super-supersign?) dissimilar systems enter into an alliance in which they offer meaning to one another and collude in the genesis of an embracing geographic icon. We have at this point a map image; but we don't have a *map* without at least title and legend and, more typically, a host of supportive signs assuming textual, pictorial, diagrammatic, and even cartographic forms. *Presentation* is the level at which the map image is integrated with and positioned in relation to relevant signs in other significant domains, and with which we have finally—or primarily—a complete and legitimized map. We will not take the position that maps are assembled from constituents (perceptually composed) or that they are dismantled into constituents (perceptually decomposed), but we will assume that the map is entered at any level of signification (perhaps many all at once), and that interpretation proceeds in either direction, by integration or disintegration, toward map or toward mark.[50] But not necessarily in a straight line. It may be tempting to regard these levels of signification—partly because of the order of their discussion, partly because of logical predisposition—as stages in a sequential process, which, set in motion, moves inexorably toward a condition of greatest or least integration. That is not our view. These interpretive levels are *simultaneous states* and, although the map—or part of a map—may occupy only one of these states at one instant for one observer, they are all equally accessible through a process of perceptual transformation—that is, a restructuring or refiguring of the map.

Elemental Signs

Elemental map signs, by definition, cannot be decomposed to yield lesser signs referring to *distinct geographic entities*. They are the least significant units that have specific reference to features, concrete (Omaha) or abstract (1,000 pigs), within the map image. Appraised in terms of the map's graphic signifiers, this criterion is easily confused; and we must keep in mind that a sign is not its expression, *but the marriage of expression and content.* The elemental map sign operates at the lower bound of the map's content taxonomy, and below this bound reside connotation and characteristic but nothing that can be construed as feature. Strict linguistic models of maps become hopelessly contorted over this issue if their analogies are pushed too far. Q.—*What is the graphic equivalent of a phoneme? A1.—There isn't one. A2.—It's a misguided question.* As we have seen, the map is an iconic medium that imposes its behavior on language, not the other way around; and there is no reason to expect graphic signs to observe the rigidly contrived, and separately evolved, protocol of phonetic representation.

At the elemental level, graphic mark (a triangular dot, a blue line) is equated with feature (an occurrence of cobalt, a river). But the elemental sign is not, of necessity, univocal. It is common practice in thematic cartography to invent map signs which (as elements) are polymorphic, polychromatic, polyscalar, and in consequence polysemic; and, although each sign generated through such principles refers to one feature, it expresses simultaneously several of that feature's attributes.[51] The elemental nature of map signs resides in the singularity of their geographic reference, not the simplicity of their meaning. Visual simplicity is no yardstick either; elemental signifiers are not restricted to visual primitives like dots and lines. They may just as easily assume more complex or more overtly iconic forms: a juxtaposition of flags signifies a border crossing, a bull's-eye a city, a string of dots and dashes a political boundary. In spite of their complexity, these are elemental signs; they are not decomposed in interpretation: one flag signifies nothing without the other; the dot of the bull's-eye cannot be stripped of its enclosing circle; the patterned line cannot be reduced to Morse Code. None of these will dissolve into autonomous signs.

The autonomy of a sign, and therefore its elemental status, can only be assessed in view of the *entire lexicon of the map that accommodates it.* Take, for example, the signification of a church with the image of a square surmounted by a crucifix. If the square is also deployed *sans* crucifix to represent buildings in general, or if other signifiers can be exchanged for the crucifix to denote a variety of building types, then the square is an elemental expression and the crucifix (or anything else) appended to it is subelemental. The crucifix is, in effect, a qualifier. Its content is characteristic, not feature; and, regardless of its symbolic potency or self-sufficiency outside the map, in the map it has no *geographic* reference independent of the square that serves as its vehicle. This is an elemental *construct,* the syntactical product of two signs, one conjugated with another. Its expression is structurally divisible into two or more signifiers with both separate and joint meaning (building + Christianity = church). If, on the other hand, the square appears only in conjunction with the crucifix, it has no reference independent of their union, and they must be jointly taken, not as construct, but as an undifferentiated element similar to the juxtaposed flags. This distinction is an important one because it indicates the presence or absence of an elemental syntax.

How are we to interpret two signifiers that apparently claim equal reference to the same feature, as both blue line and blue-tinted area do in the cartographically standard lake sign? We could regard these as coextensive signs manifest, in Klee's terms,[52] as medial and active conditions of the same visual plane. This may be valid with respect to *possible* representations of lakes, but a map can only admit one such possibility to the exclusion of all others: we will not find one lake portrayed as outline, its neighbor as colored area and the next as both.[53] Neither signifier is redundant in the map, *which adopts both*, because, in that context, neither signifies in the other's absence. An alternative analysis, equally from the Formalist perspective, would identify the lake sign as one visual element: formed by its outline and characterized by the color blue (blue in this case has no form but is only an attribute *of* form). Taken as a basis for explaining how the sign functions, how it relates content and expression, this puts us in an absurd position. A lake is

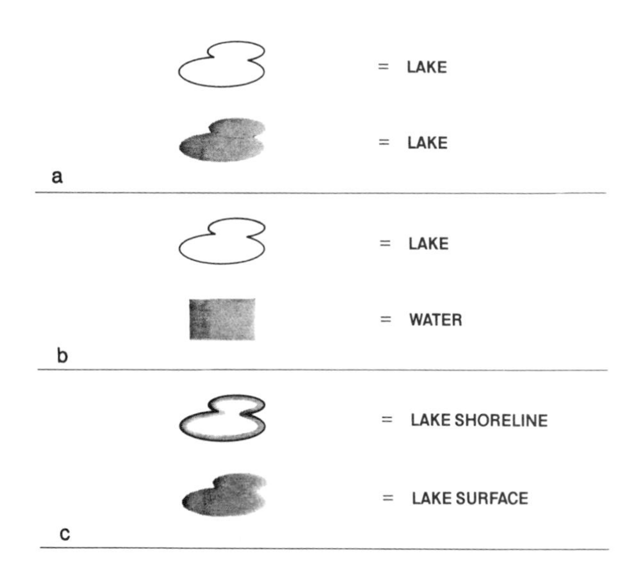

Alternative interpretations of the lake sign: *a* and *b* from a Formalist perspective, and *c* as a sign contract. The resemblance between the shoreline in *c* and pre-lithographic lake signs is anything but coincidental.

signified by a blue line that closes on itself; and, if within that figure we find a blue tint, then the lake is characterized as having water in it! Both of these postures—the former accepting line and area as simultaneous signifiers of the same signified, and the latter accepting only the line as denoting feature and denying formal status to the area it encloses—refuse to acknowledge what we already take for granted . . . that the blue line represents the shoreline of the lake and the blue tint the surface of the lake. Correctly or incorrectly, with naive or deliberate motive, this is how we interpret it, and this is how we map it. Of course the shoreline feature, strictly speaking, does not exist except as a boundary between water and land or as a locus at which the depth of the water table reaches zero with respect to the land surface (whatever that is)—and Keates' objection to the use of boundary signs in street plans applies here as well.[54] But if we can accept contour lines, and other isolines, then we have certainly learned to accept the shoreline: the *surface* of the lake is no more concrete—it is just the boundary between water and air —and the fact that it's planar (we can water ski on it) rather than linear makes it no less an abstraction.

In principle, then, we regard the land surface and the water table as roughly parallel planes (and as everywhere coextensive), and where these planes intersect, we conventionally demark their intersection with a blue line and place a blue tint to one side of that line (preferably the wet side). What we have then are two abstractions, shoreline and water surface, that we are willing to grant status as features (and to map accordingly) while at the same time recognizing them as two of many aspects of *connotations* of the lake (or pond or ocean) feature. So we have another type of sign construct (shoreline + surface = lake), only this time both of its components are features. And it turns out that the blue line, in and of itself, does not represent the shoreline after all (although it may represent a river in the same map) but does so only in the presence of a blue tint on one side and none on the other: *as part of a sign construct.*[55] Thus whereas the language of the map is drawn from a store of culturally prescribed possibilities, its terms are specifically defined only in application, where the semantic field and syntactical procedures of the individual map form a unique dialect or *sémie*.

We have tried to demonstrate why we must insist that map signs be considered in terms of both expression *and* content, and to point out the inadequacy of a Formalist perspective that regards only signifiers but not signs, as well as to suggest the degree to which our conceptualization of phenomena structures, even dictates, the manner in which we represent them. Thus an elemental sign is a *sign of elemental meaning*, one which refers to an element of the landscape that, however artificial, we are not inclined to tear into constituent bits. With this premise it is possible to build systems of signs, and systemic meaning, from elements.

Sign Systems

By sign system we mean a set or family of similar elemental signs *extensive in the space of the map image:* a distribution of statistical units, a network of channels, a matrix of areal entities, a nesting of isolines. In this respect, we identify a road system, a river system, or a system of cities. It requires that we interpret many like signs as one sign, again a syntactical product but now one of . . . *geographic* syntax. This systemic signifier is shaped by the disposition of its corresponding set of phenomena in geodesic space and by the topological transformation that brings this space to the surface of the page. It is also shaped by the way we define elements in the first place. If we were to map, say, the distribution of mountainous regions in the United States by taking as our criterion the (rather over-simplified) notion that all lands elevated 1500 meters or more qualify and that those of lesser elevation do not, we will find in our map a quite different sign system than if we had chosen 2000 meters as our benchmark. It isn't usually this innocent. What if we were mapping toxic levels of airborne pollutants? What the map says on this subject is determined by what standards, *whose* standards, we accept as a yardstick of toxicity. *In content a system is, after all, a system of features—and features only exist when we recognize them as such.*

An arrangement of signifiers on the map constitutes a system only, of course, by virtue of our ability perceptually to organize its elements into something whole. At the systemic level, the bases of

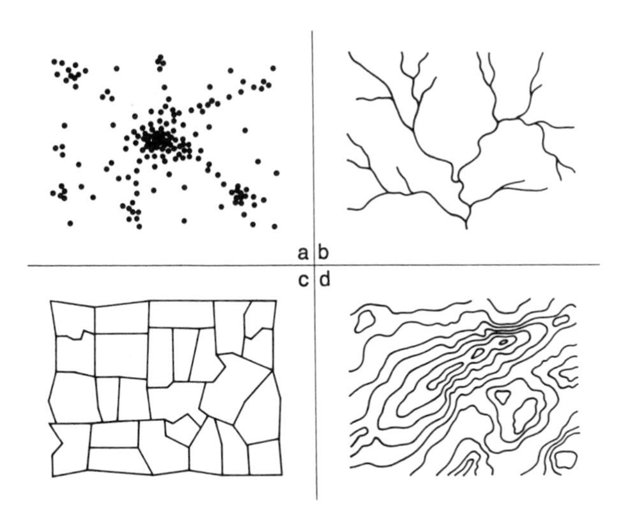

Typical cartographic sign systems: *a*, a discrete distribution; *b*, a network of signs; *c*, a sign matrix; *d*, nested signs. Regardless of implantation or graphic symbolism, each system structures the landscape in a distinctly different manner.

affinity among elements are those of *implantation* (yielding point, line, or area systems) and those formal and chromatic attributes variously termed *qualitative*, *nominal*, *distinguishing*, or *differential*. Not surprisingly, the latter are as effective among linguistic signs as among iconic signs, distinguishing hydrographic nomenclature, for example, by italic form or blue color. What is surprising, however, is the degree of variation the systemic signifier will tolerate without falling to pieces. Our highway maps, almost to the last, serve up pavement in a *smorgasbord* of colors: red, blue, yellow, black, brown, whatever is in the printer's pantry. If the object is to represent a coherent highway *system*, then we could hardly do more to subvert its recognition. But that object is secondary to the marking out of politically based *subsystems*, the sifting out of the relative accomplishments of federal, state and county treasuries. These maps can't just be written off as the products of illogical design or aesthetic insensitivity; they are graphic examples of how the extrasignificant functions of the map . . . *penetrate* to its most practical and seemingly dispassionate design decisions.

The reason we can get away with this sort of thing is that, with the exception of scattered distributions, cartographic sign systems are typified by connectivity. Their elements link up, abut, cradle or nest within one another. *They have anatomies*. We recognize primarily their structure and utilize the characteristics of their elements mainly to highlight subsystems that would be otherwise undifferentiated, or to unstick systems of similar structure. *That is to say, we attend more to the syntax of the system than the semantic import of its components*. We don't distinguish blue highways from rivers because their signifiers are a little wider and a little less sinuous, but because they are *structured differently as systems*, because they are manifestly *different landscapes*. The system is a landscape because, whereas the element simply *is* somewhere, the system . . . *goes* somewhere.

Synthesis

As we have said before, there is no such things as a monothematic map. Consider this emblem of thematic cartography: an array of graduated circles against the barest outline of subject area. Such a map image may signify a shoreline (usually elaborated beyond any conceivable utility), the water surface, the land surface, and one or more proprietary boundaries, and—almost forgot—whatever it is the graduated circles might represent. Stripping off the circles leaves us with an absolute minimum of three sign systems, and usually twice that many, lurking behind the ostensibly servile trace of the pen. Certainly cartographers design maps for cartographers—as architects design buildings for architects and politicians make laws for politicians—but to pretend that this is monothematic is . . . *insane*. Can we really take that much for granted? Are we so thoroughly hypnotized that we can't even *see* the map?

Maps are about relationships. In other words, they are about how one landscape—a landscape of roads, of rivers, of cities, government, sustenance, poison, the good life, of whatever—is positioned in relation to another. The map synthesizes these diverse landscapes, projecting them onto and into one another, with less than subtle hints that one is correlative to another or that *this* is an agent or effect of *that*. The map can't simply say that something is present (present . . . *in what?*) or that it is distributed in a certain way (distributed in relation . . . *to what?*). At this level the map image as a whole is the supersign, and the various *systems* it resolves to are its constituent signs, signs that can only have meaning in relation to other signs. Merleau-Ponty puts it this way:

> What we have learned from Saussure is that, taken singly, signs do not signify anything, and that each one of them does not so much express a meaning as mark a divergence of meaning between itself and other signs. Since the same can be said for all other signs, we may conclude that language is made of differences without terms; or more exactly, that the terms of language are engendered only by the differences which appear among them. This is a difficult idea, because common sense tells us that if term A and term B do not have any meaning at all, it is hard to see how there could be a difference of meaning between them; and that if communication really did go from the whole of the speaker's language to the whole of the hearer's language, one would have to know the language in order to learn it. But the objection is of the same kind as Zeno's paradoxes; and as they are overcome by the act of movement, it is overcome by the use of speech.[56]

What could be signified by any system of distributed dots, or branching lines, or nested lines? *Not much*. If juxtaposed with a sign system that we could recognize, or furnished with a nomenclature that allowed us to supply that system, they could become signs, not by virtue of any abstract geographic reference but *in relation to* another sign system that holds meaning for the observer.[57] If you have to resort to the map title to determine that *this* map of teenage suicides takes place in Los Angeles, then you're probably too far removed to be concerned. What the map *does* (and this is its most important internal sign function) is permit its constituent systems to open and maintain a dialogue with one another. It is obvious why a road folds back on itself when we can see the slope it ascends, or why two roads parallel one another a stone's throw apart when we can see them on opposite banks of a river, or why an interstate cramps into a tense circle when we can see the city and its rush-hour torment. We *know* the behavior of this system so well, in fact, that we can take it as an index[58] of other systems in the total absence of their direct representation. On the face of it, the map confirms these understandings; but they are understandings . . . *that have already been created by maps*.

The *gestalt*[59] of each sign system is positioned against the semiotic ground of another sign system, or a subsynthesis of systems. The roads in the state highway map aren't grounded against an insignificant white surface; they're grounded against North Carolina or Illinois or Texas. What lies between the roads isn't aether (it isn't 40 lb. Springhill Offset either): it's tobacco and loblolly pine and patches of red dirt rolling over the Piedmont, or rugose mats of corn dotted with crows and John Deeres, or relentless miles of sand and prickly pear rippling in the heat. *There is nothing in the map that fails to signify*. Not even in a map of the Moon. So the flow of water is interpreted against the ground of land form, and *vice versa;* and the pattern of forestation is interpreted against the ground of both, as both and each are interpreted against it. In the synthesized map image . . . *every sign system is potentially figure and every sign system is potentially ground.* There is nothing inherently or irrevocably ground about even the land mass: try telling a truckload of surfers the shoreline in the highway map is just a backdrop to the road system. They'll let you know you have it all backwards.[60]

The map image is a synthesis of spatially and temporally registered *gestalten*, each a synthesis in its own right; and to pretend that this whole is no more than the sum of its parts, or that we can do no more than recommend a certain alignment of their priorities, is to reduce our concept of the map to that of a diagram. No degree of thematic constriction can silence the conversation among map signs. The map models the world as an interplay of systems and presents it to us as a multi-voiced analogue, with harmonies and dissonances clearly discernible. Through the map we observe how systems respond to one another, and appraise the nature and degree of that response. We *explore the world through the map*, not as vicarious Amazon travelers hacking across the pages of *National Geographic*, but by remaking it in our own chosen terms and wringing as much meaning as we can out of what we've made.

Presentation

In presentation the map attains . . . *the level of discourse*. Its discursive form may be as simple as a single map image rendered comprehensible by the presence of title, legend, and scale; or as complex as those in *The New State of the World Atlas*,[61] hurling multiple map images, diagrams, graphs, tables, and texts at their audience in a raging polemic. It may be as diverse as vacation triptiks, rotating cardboard star finders, perspex-slabbed shopping center guides, chatty supermarket video displays, or place mats for formica diner tables. Presentation is more than placing the map image in the context of other signs; it's placing the map in the context of its audience. Robert Scholes identifies discourse, in the arena of literature, as:

> Those aspects of a text which are appraisive, evaluative, persuasive, or rhetorical, as opposed to those which simply name, locate, and recount. We also speak of "forms of discourse" as generic models for utterances of particular sorts. Both the sonnet and the medical prescription can be regarded as forms of discourse that are bound by rules which cover not only their verbal procedures but their social production and exchange as well.[62]

And he notes that the: " . . . coding of discourse is a formal strategy, a means of structuring that enables the maker of the discourse to communicate certain kinds of meaning."[63]

Discourse is preceded by a code of presentation and by the notion of an audience capable of applying that code to reach meaning *through* structure. For us, this means that the idea of "percipient" must be extended to the entire culture of mapmakers and map-users and include, as one of its most prominent aspects, their ability to generate and utilize strategic codes that permit maps to speak *about* the world rather than simply of it.

In bringing the map to this point we make it entirely accessible to the processes of extrasignification, and subject to their appropriation. It can be seized and carried off whole (necessarily whole) to serve the motives of mythic representation. The plan of the shopping center, color-coded, with shops topically and alphabetically organized and numerically keyed—a paradigm of logical graphic representation for the illogical masses—becomes an expression of the fact that "We've got it all: trendy clothes, trendy shoes, books, records, tools, cameras, jewelry, fondue pots, exotic coffees, pizza and parking." The diner placemat ceases to be a regional guide to places of interest and focal points of recreation (it was never meant as a gravy blotter or it wouldn't have been printed in the first place) and becomes the Chamber of Commerce's propaganda vehicle, complete with smiling checker-shirted fishermen tugging against smiling bass the size of Volkswagens. Which brings us back to where we started. The map is simultaneously an instrument of communication—intrasignification, given the benefit of doubt—and an instrument of persuasion—extrasignification and its propensity toward myth.

Presentation locates the map front and center in all this action, at the vertex of both planes of signification. It's not a quirk of house style that populates the *National Geographic* map with maize-laden Cherokee or the state highway map with trees, bees, civil war artifacts and cavorting tourists. It's the deliberate activation of popular visual discourse. It's not just pragmatism or objectivity that dresses the topographic map with reliability diagrams and magnetic error diagrams and multiple referencing grids, or the thematic map with the trappings of f-scaled symbols and psychometrically divided grays. It's the urge to claim the map as a scientific instrument and accrue to it all the mute credibility and faith that this demands. Presentation, as the end and the beginning of the map, closes the loop of its design. It makes the map whole and, in doing so, prepares it for a role that begins where its avowed attention to symbolism, geodesic accuracy, visual priority, and graphic organization leaves off.

It injects the map into its culture.

CHAPTER SIX

Each Sign Has a History

And the culture accepts the map. The culture receives it, is at home to it, welcomes it with open arms. A kid picks up *Winnie-the-Pooh* and makes complete sense of the map on the endpapers . . . without having had the slightest instruction in map reading. Another opens *The Hobbit* and, though the map lacks a legend, is nonetheless able to follow Bilbo and the dwarves across Wilderland. The children who read *Swallows and Amazons* experience no difficulty in understanding the relationship between Beckfoot and Holly Howe, even though the map is oriented with east at the top. Readers of *Mistress Masham's Repose* understand Raymond McGrath's map of Malplaquet, despite the fact that it's been reversed out. Although the legend to the map in *Big Tiger and Christian* omits the desert symbol, no one mistakes the pattern of dots for anything else. What kid has had a problem with the map in *Treasure Island?* With making sense of the map—and its two scales—in *Astérix le Gaulois?* With the map in *Paddle-to-the-Sea?* or the one in *Scuffy the Tugboat?*[1] Why is there so little . . . *resistance* on the part of even children to what we have just seen is an endlessly coded synthesis of sign systems, a veritable baroque of sign functions layered, one on top of the other, in dizzying density?

It is because the map is not apart from its culture but instead a *part* of its culture. It is because, as a map-immersed people, its history is our history. It is because we grow up into, effortlessly develop into, this culture . . . which is a culture of the map. Why does no kid find it difficult to make sense of the map on the endpapers of *Winnie-the-Pooh?* Because the conventions of that map are all but continuous with those *of the rest of the illustrations* in that book, and the rest of the illustrations are all but continuous with the larger *world of illustration* of which *Winnie-the-Pooh* is only a part. In its turn this *world of illustration* is seamlessly connected to a still larger *universe of representations* in which the child has been immersed since day one, a world of clowns stenciled on cribs, of stuffed animals, murals, embroidered logos, of dolls, toy trucks, pictures on cereal boxes, of cloth books, sound stories, of photos on Lego boxes, designs on wrapping paper, greeting cards, of images on Tinkertoy™ cans, records, magazine covers, of television, wallpaper, picture books, of posters, billboards, doll houses, movies. . . . The list is endless, it is a catalogue of contemporary culture . . . *and maps are a part of it*.

The map is not an alien form that came from outer space but a synthesized system of supersigns we all grew up with (that grew up with us), all of us . . . *as a people*, and each of us . . . *as individuals* (as we develop, we bring our culture into being). Of course, the use of maps is something we have to learn, as we have to learn everything in our culture (as we have to learn to speak, as we have to learn to use the toilet), but because the map is so continuous with so much of it, this is not something that is terribly hard to learn. It sort of comes with the territory: if you get much of the culture . . . *you get the map*.[2] Confronted with this map of ozone over the South Pole, with only the faintest traces of a ghostly white hinting at the continents, the rest a swirling abstraction of hot pinks and acid yellows,[3] you may find this more or less difficult to accept—but another glance at Ernest Shepard's map in *Winnie-the-Pooh* may convince you how easy it might have been . . . *at least to get started*.

And once started, the rest follows. Since the culture is whole it doesn't really matter where we start (the clown stenciled on the crib is as good a place as any). We can plunge into it anywhere (which is where we entered it as children). Because the history of the map is *our* history we are already up and running (in coming to grips with the making of maps we recapitulate this history); because the connections from the map to the rest of the culture radiate from *every* part of it, we can commence with *any* part of it (map signs come from/return to a common pool of conceptual-gestural-vocal-graphic complexes that is the cultural legacy we share). Any thread unravels everything. Here: we've pulled this one. It is that of a *single* map sign. It is that of the . . . *hillsign*, that is, the sign, supersign or sign system—sometimes the one, sometimes the other, sometimes all three layered one on top of another—that is used to commit what we know about landform relief . . . to paper.[4] The sign for river would have served as well. So would that for town or trees or roads (any would have done). But everyone agrees that hills are the *hardest* things to get right on a map. If map signs *don't* draw on the common pool, if their history is *not* our history, if they *are* weird things only a professional can understand, then we should be able to see this most readily here . . . *with this sign*. This should be the sign with which our case should be the *hardest* to make. But if we can make it here then . . . it must become obvious how the interest which society has in perpetuating itself—with its class, gender and age distinctions, its rigid hierarchy of haves and have-nots—*seeps into the map from everywhere*, even in something as apparently divorced from these concerns as an elemental map sign, as a sign for a . . . *hill*.

A Brief History of the Hillsign

Thrower says: "Delineation of the continuous three-dimensional form of the land has always been one of the most challenging problems in cartography."[5] Lynam says: "The representation of mountains has always been the map-maker's hardest problem, for mountains have length and breadth as well as height, and they hide something round every corner which must nevertheless be shown on the map."[6] Robinson and crew say: "Because of the relative importance to man of the minor landforms, their representation together with other data has been a great problem in large-scale mapping," and they treat landform representation apart from other problems in cartographic representation, noting that, "There is something about the three-dimensional (3-D) land surface that intrigues cartographers and sets it a little apart from other cartographic symbolization."[7] Greenhood speaks for the profession as a whole when he says:

> The way a map depicts relief is often a hallmark of merit "that distinguishes the handiwork of someone who knows," says Joseph T. Maddox, a geographer and also what might be called a connoisseur of maps. "Mountains are not heaped up haystacks of earth; they have length, breadth, and structure." . . . The cartographer has had to prove himself pretty clever whenever the facts he had to show upon a flat surface were themselves about flat matters, merely two-dimensional. But in showing three-dimensional facts on a flat surface he displays his true ingenuity.[8]

To the extent that it has been worked out, the historic record would seem to support these observations, though belying the glib assurance of most of these statements is the fact that little of its history is known.[9] Extant maps from periods prior to the European Middle Ages are rare from any part of the globe, and few enough have come down to us from the European Middle Ages. And as we know, the attempt to extrapolate about early map*making* from the mapping behavior of (more or less) contemporary *mapping* societies is freighted with a potentially disastrous cargo.[10] Nonetheless, it is here—with the construction of a hypothetical precursor sequence to the historical record *as such*—that any attempt to understand the lost portions of the record *must* begin.

Eskimos have been frequently singled out for their ability at relief representation, though it is becoming clear that they were not, as Bagrow once claimed, "perhaps alone in attempting the delineation of relief features."[11] In any event, these relief representations were usually *relief models* (not maps). The use of hachuring by Eskimos is also open to doubt. Though Spink and Moodie write as follows:

> Thus, in both the maps of the Caribou and coastal Eskimo, simple line-work is used to portray the significant natural features. Generally, an unbroken line is used to represent coastline or river bank, and this is backed in appropriate places by hachuring to denote the presence of cliff or mountain sides. The representation of relief is absent from many of the maps but its occasional presence is significant.[12]

But the matter is not so straightforward; for even this avowedly occasional hachuring seems likely to be a matter of interpretation, either on the part of earlier observers, or on the part of Spink and Moodie. They note themselves that its use is inconsistent (and hence may represent some other feature of the landscape) and suggest that it might not be hachuring at all:

> In some of the maps attempts are made to more accurately depict the intricacies of relief by varying both the length of the hachure strokes and their direction. The pictorial effect is encountered in charts like Figure 17b. In this area the use of extensive hachuring is quite surprising since relief on the west coast of the bay and at its head is not particularly prominent, so the lines may have been introduced merely to emphasize the trending of land, coast and rivers.[13]

Finally, they note that the hachuring appearing on some of the published Eskimo maps might have been introduced by Europeans in more than one sense:

> Native hachuring appears on few of the early charts and its later adoption may be in part a result of European influence, either by instruction or by acquaintance with the explorers' own charts. The maps collected by Captain W. E. Perry in 1822 . . . lack hachuring along part of their coasts and the hachure work which is present seems, by the regular length and frequency of the strokes, to have been produced by mechanical means, probably at the engravers.[14]

Clearly the Eskimo could build relief *models* in an amazing number of ways, and evidently they used hachure-like strokes to represent something, but despite the fact—on which all observers agree—that they were superb mappers, it is moot whether Eskimos exhibited any special facility with respect to the two-dimensional representation of relief features.

Perhaps this is because, as the linguist Raymond Gagnías claimed, the Inuit conceptualize visible phenomenon . . . two-dimensionally.[15] Or perhaps it is because they are, as Rundstrom says, *mappers*, not map*makers*: "During fieldwork in 1989, one Inuk elder told me that he had drawn detailed maps of Hiquligjuaq from memory, but he smiled and said that long ago he had thrown them away. It was the act of making them that was important, the recapitulation of environmental features, not the material objects themselves."[16] Rundstrom's anecdote, of course, is pointed. What he's saying with it is that the Inuit don't make maps . . . *even though they could*. For Rundstrom this illustrates his central tenet, that mapping behavior reflects—not potential ability (the Inuit suffer no cognitive deficit)—but cultural values and social needs.

Exactly. *Exactly*. For this is no more than to say that the development of a sign for hills reflects *not* potential ability (that had not been an issue for hundreds of thousands of years), but cultural values and social needs . . . which is to say . . . *interests*. Not until there is some *interest* in *representing* the presence of hills (the better to mark the limits of suzerainty) or the width of hills (the better to lob a cannonball over them) or the volume of hills (the more effectively to mine them) will the representation of hills emerge even in our map*making* culture. What Eskimo interest would the making of maps of hills have served? There is none in marking the limits of suzerainty, in firing cannonballs, in mining. It is hard to think of one. Navigation comes to mind, . . . *but how many of us make maps of the environments we regularly move through?* Or for that matter . . . *even use them?* New Yorkers do not leave their apartments, stand on the street with a map, and puzzle out their route to the subway. Angelenos—with far greater need—do not usually get into their car and consult a map to find their way to work. Judging from the number that stop and ask me for directions, not even long-distance truckers are big users of maps.[17] No, navigation is out. What then? Rundstrom says, "Mapping recapitulates other similar cultural behaviors, all of which spring from the basic value that the Inuit place on environmental mimicry."[18] This does facilitate navigation, but as we have already seen, although this mimicry results in *mapping*, it does not necessarily result in maps. Much less two-dimensional signs for hills. Yet the signage required to map hills is not something that just . . . *springs into being overnight*. On the contrary. As we will see, it is something that emerges *slowly* through processes of cultural elaboration. But since "storage and retrieval of hard copy were not normally a part of [Inuit] cartography,"[19] there existed no *medium* among the Inuit for this elaboration to have taken place. Were an individual Inuit to have solved the hillsigning problem—and why not? it was sooner or later solved by somebody in every mapmaking society to experience the need to represent hills—it would have been no more than an *individual* accomplishment, not a *cultural* one,[20] unless, through the transmission of "hard copy," it were to become a cultural property (were to result in the creation of a cultural tradition, in the creation of . . . *culture*). But the transmission of such "hard copy" is not an attribute of Inuit culture. This interplay between the individual and his or her culture—and between innovation and transmission—lies at the heart of what follows.

But if the representation of landform relief features is actually absent in Eskimo map*making*, then it is probably absent in the map*making* of mapping peoples generally. Few other candidates have even been put forward. P. D. A. Harvey nominates the Tewa, the Maoris, the Tuaregs and the Incas, but in each case the maps in question turn out to be models.[21] This is not to say that such peoples do (or did) not make maps, or that they can (or could) not indicate *the location* of significant landforms–merely that these landforms are (or were) not *represented* in two-dimensional form.

But if precursor representations are not to be found in *mapping* societies, they must materialize in map*making* societies, or in what might be better regarded as *proto*mapmaking societies. Many of these must have existed (at least one for every indigenous mapmaking society), but few have left traces. The Mixtec, of pre-Hispanic Mexico, did. In what form? That of unambiguous, but extremely simple, hillsigns. *Simple*, of course, is the key. Without chronology as a guide, other ordering principles must be brought into play. That relied on here is a generalization of the developmental models alluded to in Chapter Two that characterize *simpler* forms as *prior to* (that is . . . *earlier than*) more *elaborate* and (therefore) *subsequent* forms. At the same time, these simple hillsigns appear in a society where they might be expected . . . *on other grounds*. As we already know, mapmaking emerges to facilitate the control of social processes in rapidly expanding societies. It is therefore in groups exactly like the Mixtec that mapmaking *would be expected* to emerge. As we have just seen, small face-to-face mapping societies lack the *need* to *make* maps—they do not possess the values that either insist on it or make it possible—and hence fail (or refuse) to elaborate the culture to do so. On the other hand larger groups *exist as such* by virtue of, among other things, the prior development of mapmaking for purposes of social and environmental control. That is, small face-to-face groups will exhibit precisely the nascent forms of protocartography we find among Australian aborigines, the Inuit, Caroline Islanders, and the Luba,[22] whereas larger, more dynamic groups will exhibit the relatively simple map*making* that we in fact find among the early Mixtecs, Egyptians and Babylonians. (Increasingly large and complicated groups will exhibit the institutionalized *mapmaking* we find among the Romans and Han Chinese; while really huge dynamic groups, like the 19th century British and contemporary Americans, will develop cartography *per se*, the most elaborate institutionalization of the mapping impulse to date.) Thus to find in a growing and socially dynamic group like the Mixtec precisely the hillsigning forms expected is powerfully confirming.

It is difficult to understand how Bagrow could have lumped the mapmaking activities of relatively elaborate societies like the Mixtec and Aztec together with those of the Eskimos and the Marshall Islanders.[23] Thrower's mention of early Mesoamerican mapmaking in the same breath with that of early Egyptian mapmaking is much more to the point.[24] Although little of the extant corpus of indigenous Mesoamerican cartography is genuinely pre-Hispanic, much of it *is* pre-Hispanic in all but a very few details. In particular it is possible to demonstrate in the Mixtec case the remarkable extent to which maps like the *Lienzo of Zacatepec 1* observe the conventions employed in noncartographic codices predating the Conquest by as many as 200 years.[25] Among these conventions is a partial system of logographic writing in which certain forms "are not merely pictures, but logograms—signs that represent one or more words in the Mixtec language."[26] One of these logograms means "hill." Smith describes it as follows:

> The sign for the Mixtec word *yucu* or "hill" is essentially a conventionalized "picture" of a hill. It is usually a green or brown bell-shaped form on a base that consists of a narrow red or blue band below which there is often a yellow scalloped border. At times the lower corners of the hillsign curl inward, forming volutes on either side. Often the outline of the hill shape is broken by small curvilinear or rectilinear projections which indicate the roughness or "bumpiness" of the hill. The hill sign has many variant shapes. For example, one side of the hill may be extended in a manner that suggests a slope and at times this extended slope functions as a platform for human figures.[27]

Logograms like this for *yucu* were used in two readily differentiated fashions. On the one hand they were used much as we use words, to name a place, and it is in this fashion that they appear in the historical narratives of the codices. On the other hand they were used much as we use a combination of words and symbols on a map, to identify *and* locate a place, and it is largely in this fashion that they appear on the Mixtec maps. I say "largely" because their use on these maps was really more complex:

> Zacatepec 1 is both a cartographic and a genealogical-historical document. The boundaries of Santa Maria Zacatepec are defined by a large rectangle that encloses all but the top register of the Lienzo. Attached to the border of this rectangle are the signs of the names of the boundary sites . . . The historical-genealogical narrative begins in the upper-left corner, extends across the top to the upper-right corner, and

> then continues within the rectangle, where it is organized in a rather rambling meander pattern. In addition to the principle narrative, which is connected by roads and warpaths, the large rectangle formed by the boundaries contains three types of place signs: (1) "non-cartographic signs"—that is,signs of towns which are actually located outside of Zacatepec's boundaries but which are placed *within* the rectangle of boundaries in the Lienzo, (2) the signs of Zacatepec's *estancias* or subjects, and (3) signs of uninhabited geographical features such as hills and rivers.[28]

Immediately you can sense the flux in the representational system—linguistic and pictorial, narrative and cartographic—but also the powerful . . . *instrumentality*. At this early stage in the development of landform relief representation, not only were signs originally developed as names for use in narratives being adapted as "pictures" to be used on maps, but they were being used to mark the boundaries of the places ruled by the people named in the genealogies to which they were attached. At this early stage there was no question of connecting these people *through* the map to other aspects of a vast system embodied in oaths, pledges, rituals, codes, laws, ledgers, and contracts (as through the map our ownership of Lot 126 is connected to the privileges and obligations its ownership entails). There was no other system, it was all here, whole at this point, *map-history, spacetime* . . . teetering on the edge of splintering into the endlessly filiated systems to come.[29] But, then, in the map . . . *still* the linkage of two sets of names . . . *the names of places* (and do not overlook the instrumentality implicit even in these, since, as Harley reminds us, "naming the land is one of the most emotive and symbolic acts that cartography constructs"[30]) and *the names of rulers* . . . legitimated in their suzerainty through the anciency of the claim substantiated in the genealogy. Making all this possible, facilitating the naming *and* the linking . . . the hill*sign* (among others) . . . bringing into the human world through the act of naming . . . *the hill* . . . and then using the hill—now captured, now separated from the rest of the world with which it had hitherto been whole—to bind space to the exercise of human will. And every map sign does this.

And yet in these early Mixtec maps, the transition to map sign was not complete and in Smith's words, "Its intent is to represent the *name* of the hill, not the hill itself; and in the typical pre-Conquest manner, the interior of this hill sign contains no shading nor any elements, such as foliage, which are not meaningful in terms of the hill's name. The place sign is not a generalized portrait of a hill based on perception; it is a pictorial sign that reflects language rather than landscape."[31] That is, the place sign is tied to the land *only through the medium of language*.[32] However, it is in this manner that hillforms first

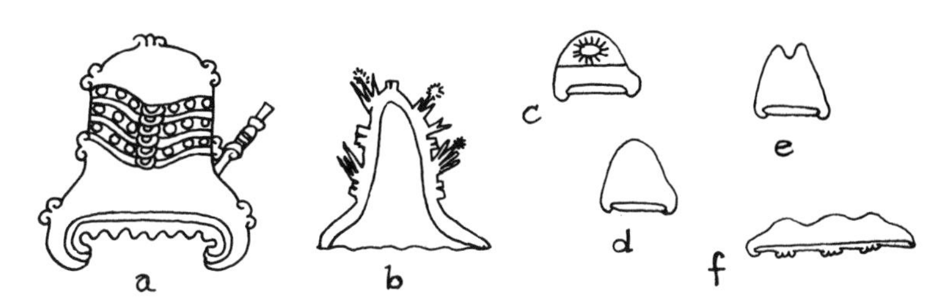

Mixtec and Nahuatl hillsigns: *a*, Codex Nuttal; *b*, Lienzo of Zacatapec 1; *c–f*, Codex Tepetloaxtoc.

found their way onto maps (it is in this way that maps first came into being), driven, in this case, by the roiled complexities of land ownership amid dynastic turmoil.[33]

It is unknown how the Mixtec or the Aztec would have continued this development for it was at precisely this point that the Conquest occurred. It is, however, possible to see how the hillsign was generalized under Hispanic influence, as in *Lienzo Zacatepec 2*, which was produced only two generations after the earlier version; or as in that part of the Codex Tepetlaoztoc reproduced in Bagrow.[34] In the latter it is very obvious that the logogram was simply generalized into a generic hillsign, and it appears in four variants: (1) as a placesign (with the additional logographic signs needed to "spell-out" the proper name), (2) in identical form but as a generic hill, (3) modified to indicate peculiarities of shapes in hills, and (4) multiplied into a range of hills, or perhaps mountains.

The similarity of this latter (multiplied) form of the hillsign to that on the well-known clay map from Nuzi in northern Mesopotamia[35] is sufficiently striking to suggest that a similar process of cartographic signmaking likely unfolded in Mesopotamia via early Sumerian ideograms, and conceivably in all map-making civilizations in which landform relief is represented. *That is, driven by the need to keep records, early cartographic signmaking probably developed conjunctively with early linguistic sign-making (and the first phase of urbanism), writing and mapping at first both growing together, sometimes difficult to completely separate, but subsequently following increasing distinctive routes, writing moving toward history and descriptive narrative, cartographic signmaking toward painting and maps.*[36] This fusion is clear in the Mixtec and Nahuatl cases, and the proximity of the approximate dates assigned the Nuzi map (c. 3800 B.C.) and the development of Sumerian ideograms (c. 3500 B.C.) is similarly suggestive.[37] Weaker cases along the similar lines might be made for Egypt and China.

Whatever the *origins* of landform relief signs in the West—name, boundary, territorial marker—their character was little changed in the 4,000 or so years following their appearance on the Nuzi map.[38] Indeed, there was little impetus for them to. Those that show up on the *Tabula Peutingeriana* (c. A.D. 500) are more than kissing cousins, and no really notable changes occur until the later Middle Ages. Speaking of the cartography of the 12th and 13th centuries, J. K. Wright observes:

> Symbols representing the various features of the earth's surface were more or less conventionalized, though we can hardly say that any definitely developed "conventional signs" were in use. . . . On medieval maps such elements as mountains, forests and cities were shown as they appear from the side. . . . Mountain ranges were generally represented by jagged, sawtooth lines running parallel to straight lines; particularly high or famous peaks by a single great pyramid. Such pyramids are prominent features in the Beatus series . . . .[39]

Such (undoubtedly conventional) signs are in all essentials those that came into being with writing 4,000 or 5,000 years earlier under the aegis of early urbanism. With the rise of the mercantile city and the gradual onset of industrialization, however, hillsigns begin to develop rapidly, and their history from this point forward has been frequently summarized. In Lynam's account, covering the period from about 1250 to 1800, the essential changes involve a gradual shift from an elevation or profile view ("rather like cock's combs"), through an oblique or bird's-eye view ("little rows of shady sugar-loaves"), to the use of the plan view (leading in the 18th century to the hairy caterpillars "found crawling across maps of Asia and America until the end of the 19th century"). Driven by the needs of the military and mining communities, and by the internal dynamics of an increasingly professionalized mapmaking itself (an instrumentality no more innocent than that of maiming and mining), this shift in perspective was paralleled by the development of conventionalized shading, from the arbitrary medieval practices of shading profile views, through the "obliquely" and usually eastern shading of later bird's-eye views, to veritable vertical shading of plan views.[40] This led, in Skelton's view, to the development of hachuring:

> Early in the eighteenth century, cartographers began to draw their hill-hatching as if vertically shaded or illuminated from a source above the object. From this method, which facilitated the representation of relief features in plan, developed hachuring by parallel lines drawn in the direction of the slope, the steepness being indicated by the thickness of the hachuring and the interval between them. This convention was used with plastic effect in 1757 in the physical maps of Philippe Buache.[41]

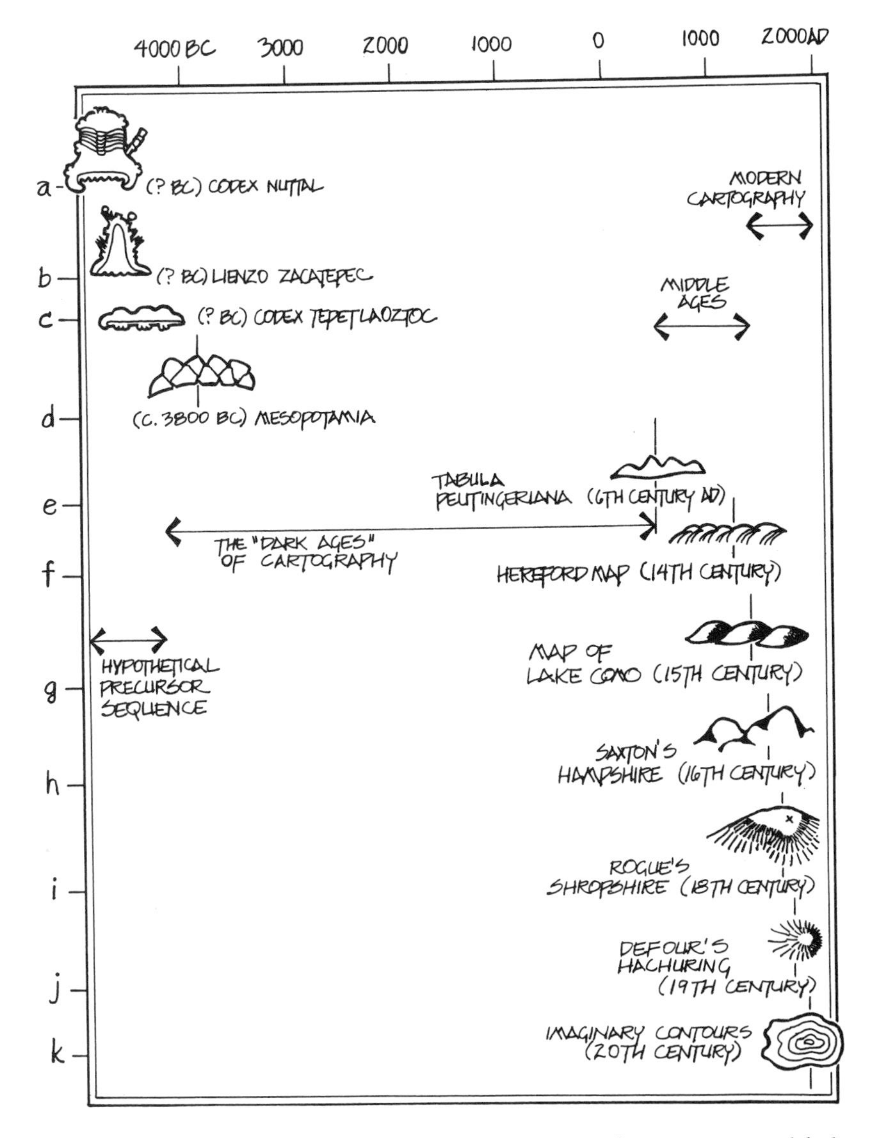

The ethnogenesis of hillsigns. The hypothetical precursor is what we saw exemplified in Mixtec and Aztec.

But Buache had already used contours on maps 20 years earlier (1737), and by the time hachuring became adequately refined (in the 1799 work of Lehmann), it was already being supplanted by the use of this still more abstract convention. Still, it took most of the 19th century for the contour to establish its superior ability to embody the interests—in naming, claiming, marking bounding, crossing, climbing, mapping, mining—accumulating around the hillsign.[42]

On small-scale maps, the use of contours resulted in layer-tinted relief representation, both with and without shading. A number of other techniques have since been propounded, but as Robinson and Sale once pointed out, "Most of them are relatively complex and intellectually involved. Their use is limited to the professional geographer and geomorphologist, whose knowledge of landforms is sufficient to interpret them."[43] Of course this was once said of layer tinting, contours and hachuring—probably of all innovations in relief representation—but Robinson and Sale were not implying that the history of relief representation had come to an end. In fact, as they looked into the future, they suggested the opposite, that it has along way to go: "For many years to come the representation of land form on maps will be an interesting and challenging problem, since it is unlikely that convention, tradition or the paralysis of standardization will take any great hold on this aspect of cartographic symbolization."[44] This seems especially likely in view of the fact that the full panoply of historically developed types is currently in wide use. If the plan view and the contour have taken over the large scale topographic survey, the bird's-eye view and hachuring are very much in evidence in physiographic diagrams, and landform and perspective maps (despite Raisz' caution that his tachographic symbols "not be placed so regularly as to look like fish scales", they still look like they were nurtured on a small-scale map of the 16th century[45]). Even simpler hill-signs are in wide use. On a recently produced map of Los Angeles created for Japanese tourists, the hillsign, though clearly derived from the tradition of the Japanese woodcut (with its obvious "special interest"), rustles back through late medieval woodcuts (with their "mountains portrayed as enormous overlapping slabs of rock" in Lynam's words) to the very earliest attempts at portraying relief.[46] These and other historic forms thrive among us.

Hillsigning Among Contemporary Americans

If professional cartographers regard relief representation as a challenge, the average American regards it as something more . . . or less. In a sample of 2,050 experimental sketch maps pulled from my collection representing a diversity of mappers, terrains and scales, only 157 (less than 8%) represent relief *at all* (although a few others have "hills" or "mountains" written in appropriate locations). When represented, relief usually consists of schematic, "oblique" views of mountain ranges at the state (North Carolina, Puerto Rico) or smaller scale (the United States, Latin America, the world). Relief shown at larger scales invariably takes the

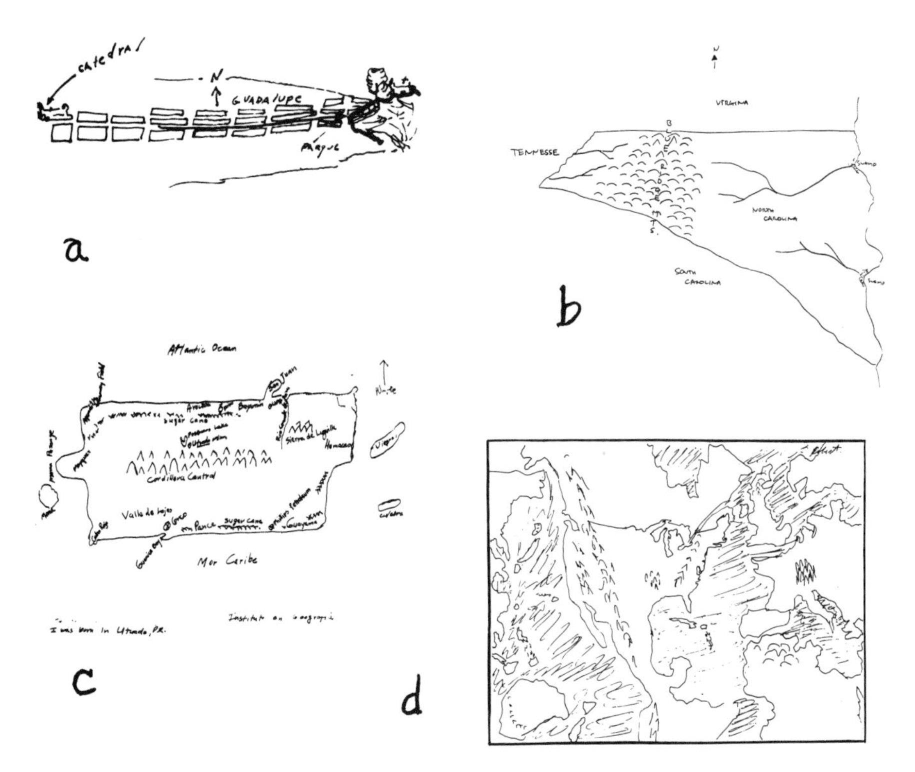

Some hillsigns from sketch maps: *a*, an oblique "picture" on a large-scale sketch of a barrio in San Cristobal las Casas; *b*, North Carolina as drawn by a Kansas college student; *c*, Puerto Rico from the pen of a native secondary-school teacher; *d*, the world by an adult resident of Connecticut.

form of oblique *pictures*.[47] A review of the few experimental sketch maps that have been published by others confirms these findings—few examples of relief at all and those highly schematic at the state or smaller scale—with the exception of a set of "drawings" of the San Fernando Valley collected from first, third, and fourth graders by Klett and Alpaugh. Of the 10 drawings they reproduce, five include obvious representations of relief: three of these are schematic obliques, one a shaded oblique and one a schematic profile.[48] In general, few cartographically naive mappers volunteer representations of landform relief. Those that do, do so at relatively small scales and employ highly schematic bird's-eye views of mountain ranges, that is, they map hills like Babylonians or medieval scribes.

Yet we cannot conclude from this that they are not aware of or uninterested in landform relief. In an ongoing study looking at the roles played by landforms in the lives of children, I have demonstrated that they're acutely aware of slope. Kids actively exploit it in bicycling, skate-boarding, skating, sliding, sledding, rolling, running, jumping, flying, *et cetera et cetera*.[49] When they cycle, they always try to maximize the downhills (many claim to succeed). Riding or walking adults do the same. Where, in cognitive image studies employing both verbal and graphic interrogatories, hills are infrequently *drawn*, they are frequently *written about*. In my own work in San Cristobal las Casas, Chiapas, Mexico, an important hill in town was the twenty-first most frequently *drawn*, but the twelfth most frequently *mentioned* landscape feature; and a similar ratio appeared in cognitive image studies in the same town in individual neighborhoods in which hills were prominent.[50] Maps of their neighborhoods drawn by a hundred high school students in Worcester, Massachusetts—a town known for its "seven hills"—frequently included the *word* "hill," often in neighborhood names (as in "Grafton Hill"), but lacked representations of relief.[51] Such examples could be multiplied indefinitely. The fact is, kids and adults know about, talk about, and write about hills and other forms of landform relief, but they don't map them.

Doesn't this demolish the point we're trying to make? Not at all. The claim is not that everyone *makes* maps—in a world of labor specialists this would be redundant[52]—but merely that the *conventions employed* in mapmaking—the signs, supersigns, and sign systems—are continuous with the rest of the culture. Why then don't children and cartographically naive adults map hills? Probably because they haven't spent their lives wrapped up in the attempt to represent landform relief *together with* other landscape attributes. *This* is the problem:

> If he shows the surface in sufficient detail to satisfy their local significance, then the problem arises of how to present the other map data. On the other hand, if the cartographer shows with relative thoroughness the nonlandform data, which may be more important to the specific objectives of the map, he may be reduced merely to suggesting the land surface.[53]

But if this is the problem, what if it were simplified, and instead of having people show *all* that, they only had to . . . *draw a hill?*

To answer this question we collected 500 drawings of hills from 300 people between 3 and 30 years of age. All North Carolinians, they lived in each of the state's three distinctive physiographic regions (coast, Piedmont, mountains). Whereas the youngest kids were simply asked to draw a picture of a hill (we did record the stories they told about the drawings), we asked the older kids and adults *either* (1) to draw profiles, obliques and plans in that order; or, (2) to draw a hill, and then to draw it from perspectives other than the one they had used first.[54] For each drawing we noted (1) the point of view from which it was made (profile

School age

	No.	Shape	1 Pre-school n=50 %	2 Kindergarten 28 %	3 1st grade 30 %	4 3rd grade 26 %	5 5th grade 25 %	6 9th grade 29 %	7 10th grade 44 %	8 College sophomores 38 %	9 Graduate students 30 %
Elevation	1		16								
	2		4								
	3		6	7							
	4		38								
	5		14								
	6		4	11			4				
	7		42	50	7	8		24	28	16	3
	8		6	14				7		18	
	9			61	50	15	72	27	38	42	36
	10			7	30		4				
	11		8					10		5	
	12				7	23		7	14	8	
	13				10			7		8	7
	14		8		20	27	8			3	
	15				13	49	84	17	10		
	16							3	2		
	17	Other		4						3	
Oblique	18								10	13	20
	19				7	4				3	7
	20					8	16	10	14		
	21								2		
	22	Other									
Plan view	23							17	19		
	24							14	24		10
	25							14	12		7
	26								5	8	
	27								14	11	
	28								2	3	3
	29							3	12	71	40
	30							14	2	3	
	31	Other							2	3	7

Ontogenesis of hillsigns. School age is displayed along the horizontal axis; hillsign type on the vertical.

view, oblique or plan), (2) the hill's form (mare's nest, breadloaf, gum drop, Alpine, whatever) and (3) the number of hills drawn. These observations together with the number of draftsmen in each age group and the percentage drawing a given hill type are shown here. (The columns don't sum to 100% since two-thirds of the respondents drew two or more hills.)

There is a straightforward relationship between age and both (1) point of view and (2) the range of hill types drawn. That is, with increasing age, there is both (1) an increasing likelihood of representation in plan and (2) an increased repertoire of representational types. Some kids chose not to draw hills *per se* but to score the significance to hillness of *slope* (as opposed to, say, *earth*) by drawing things like roller-coasters and roofs. A smaller number of kids, confined to the youngest, drew animate hills (one had circles and rays representing "the eyes of the hill"). When the very young kids were explicitly requested to draw hills from above, they responded with more hills like those they had already drawn (which is not to say that these kids could not have been taught to draw hills in plan, just that they hadn't yet discovered [learned] how as yet[55]). In general, the data are a good indication of the range of hill-form types available to and used by most Americans.

The Sequence in Which Children Acquire Hillsigns Parallels That in Which They Were Acquired in Our History of Mapmaking

There is thus a striking parallel between the development of hillsigns in contemporary Americans and the development of hillsigns in the history of mapmaking. This parallel is more than an artifact of similarly constructed charts. It is more than apparent. It is real. In both developmental sequences the first hillsign to appear, early in an individual life (and extremely early in cartographic history), is a concrete picture of a generic "anyhill" represented in elevation, egocentrically, as seen by the frontally oriented erect human being. The second hillsign to materialize, between ages 10 and 15 (and during the Middle Ages), is either a foreshortened version of the "anyhill" in elevation, or an abstraction based on its shadow-throwing properties. In either case, hills are now differentiated into classes (like rolling, isolated or mountainous)

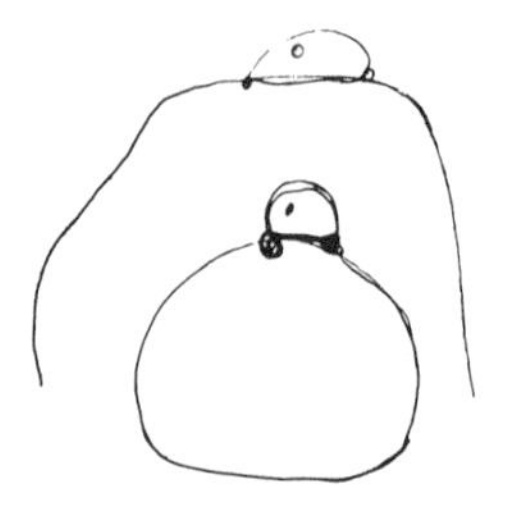

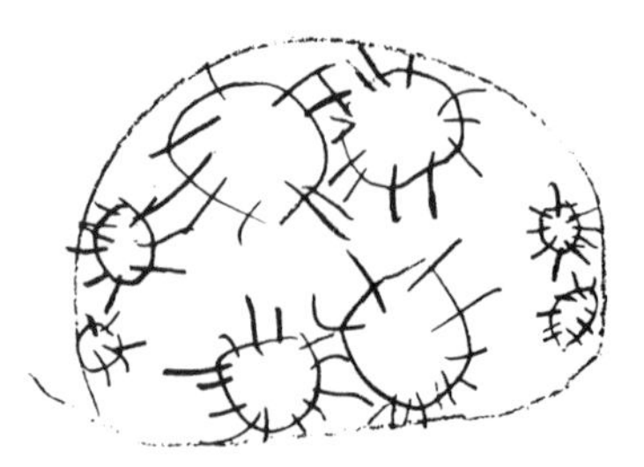

"Animate" hills drawn by a preschool and a kindergarten child.

and shown in oblique, as if from the perspective of a bird. Finally, the hill comes to be represented by contours, abstractions founded on the abstractions of a basal datum and constant elevation. In this phase hills are revealed in plan, as if seen from directly overhead, in all their unique individuality. In both sequences the pool of available hillsigns, initially small and unorganized, broadens to embrace the totality developed by "bringing forward" each type generated, *so that higher stages of development do not so much supplant as subsume lower ones in a structure of increasingly hierarchic integration*, perpetually subject to superordinate considerations of intentions to communicate, record or analyze. The parallelism of these sequences raises two questions: why either follows the order observed, and what—if any—relationship exists between them.

The Mastery of Hillsigning in Contemporary Kids

In our mapmaking society, 3-year-olds spontaneously draw pictures of hills. In many cases they have been able to recognize, point to, and name such drawings since age 2 or even earlier. Here are a few examples of hills drawn at home by my oldest son under typical, middle-class conditions *in response only to inner urgings*.[56] In the first example, drawn at 3 years, 2 months, the circular scribblings were declared "fountains," the tall vertical lines crossing them "high hills," and the pointed figures along the bottom "mountains." It is similar in several respects to those collected from other preschoolers—ages 4 through 6—in response *to requests* for drawings of hills (one of which is also illustrated here). The next example, drawn at 3 years, 3 months, was characterized as "two mountains, three mountains." Here the montiform shapes are obvious. A third drawing, produced shortly after "two mountains, three mountains," was described as "hills and letters." These drawings made by one child within a 40-day period reveal something of the lability genuinely associated with representation at this age, but they also exhibit the effects of the media in which they were produced.[57] The first was drawn with ballpoint on sketch paper, the second with art-marker on shirt cardboard, and the third with magic marker on newsprint. Incidental marks occasioned by the muscular effort required by the use of ballpoint on uncoated papers can override a child's original intentions. Such marks are less likely in easier media like pencil on paper. Other media are still less exacting: here, this is a hill *torn* from paper by the same child at 2 years, 8 months. (Children working in salt trays—where they draw directly in a thin film of salt—produce representations up to 6 months earlier than they can in other media.)

The range of hill forms drawn by this single child can be amplified by looking at a small sample from the 73 hills collected from the

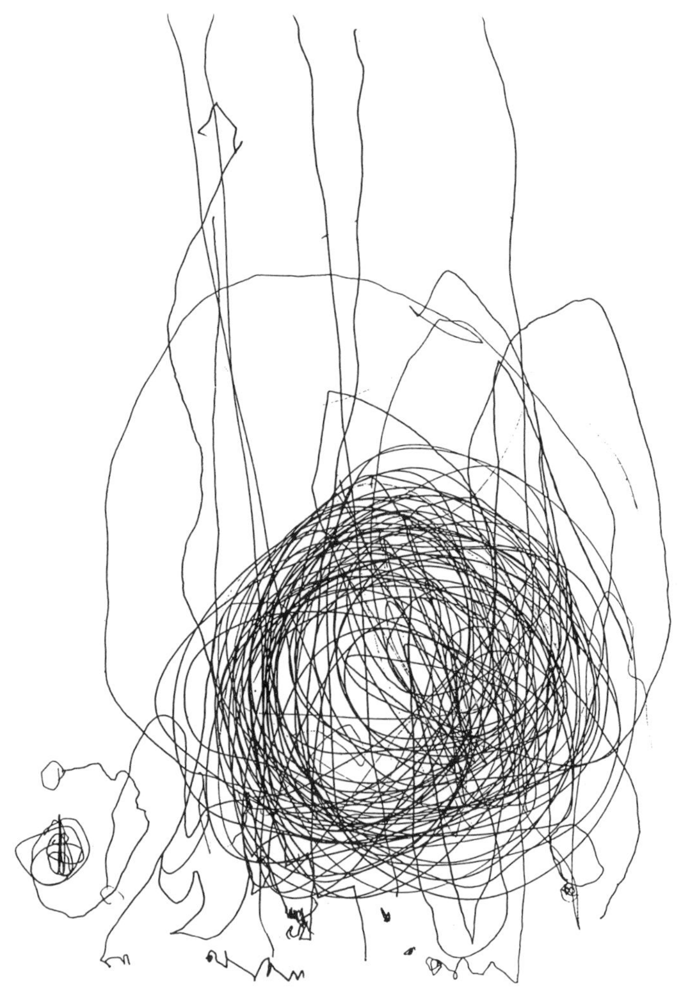

Randall's first drawing of hills—in ballpoint on sketch books paper, age 3 years 2 months.

preschoolers alluded to above: shapes run the gamut from a crude "rat's nest," through various "lumps," to what can only be described as hills and mountains. Loose ovals lying with their long axis parallel to the bottom of the paper, and inverted Us predominate, and these differ from each other only in the degree of basal closure. Other variations play on steepness of slope and sharpness of peak. Although some of the hills are animate ("the eyes of the hill," a sheeted ghost) and others are not, this is not, except as it underscores the *thingness* of the hill, a salient

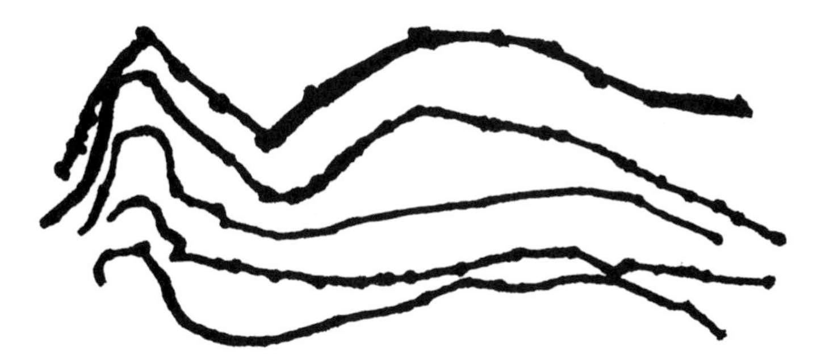

This is "two mountains, three mountains," redrawn from Randall's marker on short cardboard, effort of 3 years 3 months.

Randall's "hills with letters," of 3 years 3 months, but later than "two mountains, three mountains." Redrawn and reduced (as are all these drawings of Randall's) from the original marker-on-newsprint drawing.

dimension. Young children frequently animate all sorts of things, things like the sun or clouds, houses or cars, but rarely *non*things like ground or sky, river or rain (though they might animate individual raindrops). *Non*things have a pervasive background quality, undeniably existent, but impalpable, unquestionably representable, but typically insusceptible of animation. Things like the sun and the clouds have a foreground quality, concrete, disaggregate and manipulable. Hills, at least at this stage, fall

Hill torn from paper by Randall at 2 years 8 months.

into both camps: which depends on the momentary role occupied by the hill in the mental world of the child.

In addition to the thingness of these *single* hills (a trait dominating the drawings through the first and second grades when *groups* of otherwise similar hills begin to gain in favor), they are characterized by the perspective from which they're conceived and drawn. Each, excepting the crudest forms, is flat and seems to be drawn in elevation, as a profile or even a silhouette, taking the form of a distant hill seen by the standing child . . . essentially *like the child*, made in his or her . . . *own image*, from his or her own best perspective, erect, independent. The whole hill—solitary, concrete, undifferentiable, upright and profiled—would seem explicable, indeed predictable, taking into account only structuralist models of child development[58] and the essential aspects of hillness, which, in the Indo-European world at any rate, embrace *prominence*, *projection and jutting out*.[59]

The only problem with this scheme is that young children typically do not encounter their first hills as distant profiles but as mammoth RISINGS-UP in front of them or FALLINGS-AWAY beneath their feet. The typical environment of most of those drawing for us, even those from the mountains or country-side, is *too close up* to permit the sightings needed to experientially undergird the simple structuralist model.[60] The hill for the Piedmont urbanite (our largest group) is usually *underfoot* where it facilitates or impedes walking or running, riding or sliding, but not the construction of the hillsigns found. The veritable hills of the child's experience are multiform, complex,

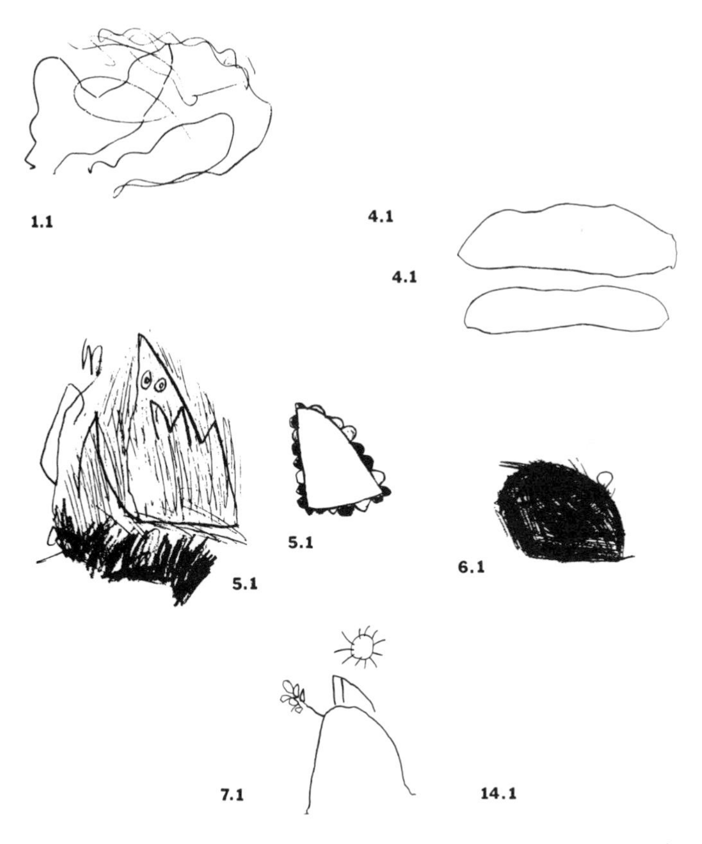

Examples of hills drawn by preschoolers. The first number in the code refers to the hilltype on the chart displaying the ontogenesis of hillsigns; the second number refers to the age group.

covered with trees, houses and other buildings, joined by bridges, disguised by cuts and fills, and nearly never isolated. They are practically . . . *impossible to see* in profile. Instead, they are encountered as radically foreshortened slopes perceived in practically every mode . . . *except the profile*.[61] Nor do the hillocks constructed in sand and dirt play provide a model for the profile view, for these are typically viewed in oblique. A profile view could only be had lying with the chin in the sand—or, assuming rotation, the temple—positions rarely seen in hundreds of hours of observation.[62] Leaf mounds, piled snow, dumps of sand or fill could play the needed role, except that these seem to be *precharacterized as hills* by children as young as age 3, suggesting that they *conform to* an existent image rather than laying the groundwork for it.

It seems likely that children are simply *taught* to see and represent hills in profile; and that the resultant image acts as an armature around which subsequent experience—including schooling—is wound and with which it interacts. There is no implication here that structuralist notions of development should be thrown away, but rather that these principles—intuitively grasped by sentient adults through unselfconscious introspection of their own cognitive processes—form the bases of their efforts to educate their children, efforts, it must be stressed, that focus on the hill incidentally if at all. For example, take the bedtime reading of *The Little Red Caboose*, a book by Marian Potter with pictures by Tibor Gergely which has sold over a million copies in English alone since 1953, has been frequently translated, and is widely available in a number of formats.[63] The story's hero, the little red caboose, is upset because nobody waves at him. Nobody waves at him because he always comes last. This always-coming-last ultimately gives him an opportunity to prove his worth by saving the train. This encourages children to save their biggest waves for him. In addition to its moral and ethical messages, the story carries information about order (*first then last*) and direction (especially *up and down*). Although the first half of the book *does* show the train going up and down and through and around any number of hills—in back of which profile views of distant relief decorate the

The train goes up the mountain. Note profile of hill sweeping across page, others in background. (From Marian Potter, *The Little Red Caboose*, illustrated by Tibor Gergely. Copyright 1953 by Western Publishing Company, Inc. Used by permission.)

horizon—the second half of the book brings a mountain, and its defining characteristics, into the very foreground of the story:

> One day the train started up a mountain.
> Up went the big black engine.
> Up went the box cars.
> Up went the oil cars.
> Up went the coal cars.
> Up went the flat cars.
> Up went the little red caboose.[64]

As the reader's voice and arm, rising with every iteration, relentlessly emphasize the direction taken, the illustration forever weds it to the mountain. To the left, the train crests a high peak which falls sharply to a vale at lower right. Here the tracks recede into a tunnel under a lower, but still significant, mountain. In the background, a distant town climbs a high hill to a castle, behind which yet other hills raise their summits to the sky. All, including the greater part of the mountain ascended by the train, are seen in profile against the chanting of the narrator's, "Up . . . Up . . . Up. . . . " Suddenly:

> "Look out, little caboose!" called the flat car.
> "The train is starting to slip back down this long tall mountain!"
> "Not if I can help it!" said the little red caboose.
> And he slammed on his brakes.
> And he held tight to the tracks.
> And he kept that train from sliding down the mountain![65]

While the reader's gestures mime the uncontrollable descent of the train,the pitch of his or her voice drops emphatically on each of the "downs." At the same time, minimal foreshortening in the illustration sends the train vertically UP the page, heightening the DOWN of its potential crash, while simultaneously permitting the slopes to be rendered in what can only be called a profile view. The story is powerful: the most blasé aunt or uncle gets caught up in the excitement, swept along by Potter's prosodic devices and the simple effectiveness of Gergely's illustrations. On the child the combined impact of text, heightened and inflected vocal rendering, supportive gestures and illustrations is tremendous, especially after 50 or 60 repetitions!

Watty Piper's *The Little Engine That Could* is another, even more popular case in point.[66] After unsuccessfully begging several other engines to replace their defunct one for the trip across the mountain, the dolls and toy animals convince a little blue engine to help them:

> "But we must get over the mountain before the children awake," said all the dolls and toys.

The very little engine looked up and saw the tears in the dolls' eyes. And she thought of the good little boys and girls on the other side of the mountain who would not have any toys or good food unless she helped.

Then she said, "I think I can. I think I can. I think I can." And she hitched herself to the little train. . .

Up, up, up. Faster and faster and faster the little engine climbed, until at last they reached the top of the mountain.

Down in the valley lay the city. . .

And the Little Blue Engine smiled and seemed to say as she puffed steadily down the mountain.

"I thought I could. I thought I could. I thought I could. I thought I could.

I thought I could.

I thought I could."[67]

Even the layout of the text makes the point! If the threat in *The Little Red Caboose* was that the mountain would force the train back DOWN, here it's that it will prevent it every getting UP. The mountain looms, a towering prominence prohibiting travel from valley to valley. In the critical pages of the actual crossing, the mountain, gently rounded, is shown in profile with the train track on its upper surface just like in a kid's drawing of something on a hill (which, in effect, it is). On the first three of these pages the train climbs from left to right; on the next page, still traveling in the same direction, it pulls over the crest; on the final two it descends, still moving left to right. The pages can be spliced together to form a single image up, up, up, then down, down, down. When read aloud the stew of text, inflected voice, supportive gestures,

Another train goes up a mountain. Note that the train rides "on" the ridge line; we can see "under" the cars. (From *The Little Engine That Could*, illustration by George and Doris Hauman, retold by Watty Piper. Copyright 1954 and 1982 by Platt & Munk Company, Inc. Used by permission.)

illustrations, and even typographic layout powerfully render the archetypic image of the hill or mountain.

Hills play important roles in surprisingly large numbers of children's books. Typically they are gently rounded prominences shown in profile, as in Wanda Gag's classic *Millions of Cats*.[68] After the old couple decide to get a cat, the old man, ". . .set out over the hills to look for one. He climbed over the sunny hills. He trudged through the cool valleys. He walked a long, long time and at last he came to a hill which was quite covered with cats."[69] As far as hills are concerned it is the illustrations in Gag's work that carry the brunt of the load, though even here the text reinforces their up-and-down character. Particularly evocative is the picture of the old lady coming down the side of a hill whose slope is mirrored in the house smoke, the clouds, and the layout of the text. Hills appear in their quintessential form in the work of another popular author, Virginia Lee Burton, though rarely without a hint of foreshortening. In *Choo Choo* she describes the train's route as, "Through the tunnel and over the hills/Down the hills, across the drawbridge,"[70] though so effective is her illustration that she needn't have bothered. The Little House in her famous *The Little House* stands on a hill initially presented in profile, but with increasing obliquity in subsequent illustrations until the end, when it again appears in profile.[71] In these pictures of rural quietude that open and close the book, the house on the hill comprises a unitary image of immense attraction for most readers. The same hills populate Burton's *Mike Mulligan and His Steam Shovel*,[72]

Smoke, clouds, type—all reinforce the shape of the hill. (From Wanda Gag, *Millions of Cats*. Copyright 1928 by Coward-McCann, Inc. Copyright renewed 1956 by Robert Janssen. Used by permission of Coward, McCann, & Geoghegan.)

Still another train climbs hills. These begin to suggest an oblique view, but the profiled crest cuts the sky in precisely the same way as did the Little Engine That Could. (From Virginia Lee Burton, *Choo Choo*, Houghton Mifflin, 1937. Copyright 1937 by Virginia Lee Burton, 1965 by Virginia Lee Demetrios. Used by permission of Houghton Mifflin.)

though here they are cut, flattened and filled for canals, passes and airports. It was also Mike Mulligan "who lowered the hills and straightened the curves to make the long highways for the automobiles."[73] The trees, houses and telephone poles in the accompanying illustration stand perpendicular to the surface of the profiled hills (rather than parallel with the edges of the paper)—just as in kids' drawings. When Mike and his shovel leave the city, "They crawled along slowly up the hills and down the hills till they came to the little town of Popperville."[74] The illustration invites the reader—child or adult—to trace the route with a finger, UP and DOWN each of the hills.

Hills are also defined far more incidentally, the cumulative impact of which could be far greater than from the more direct descriptions already mentioned:

> Jack and Jill went up the hill
> To fetch a pail of water;
> Jack fell down and cracked his crown,
> And Jill came tumbling after.

Clouds, tree form, rows of lowers, and type all echo the hill's shape. This is repeated page after page. (From Virginia Lee Burton, *The Little House*, Houghton Mifflin, Boston, 1942. Copyright 1942 by Virginia Lee Demetrios, 1969 by George Demetrios. Used by permission of Houghton Mifflin.)

The hills after Mike Mulligan and Mary Ann had cut the roads through them. Note the houses, trees, poles—all are orthogonal to the surface of the hills. (From Virginia Lee Burton, *Mike Mulligan and His Steam Shovel*, Houghton Mifflin, Boston, 1939. Copyright 1939 and 1967 by Virginia Lee Demetrios. Used by permission of Houghton Mifflin.)

Even *sans* illustrations the relation of *hill* to *up* and *down* is hammered home; and the most typical illustration, showing Jack and Jill ascending one side and tumbling down the other, shows a hill in profile symmetrically arrayed around a well. The same form can emerge in the absence of any hint of a hill. A typical illustration for:

> Goosey, goosey, gander,
> Whither shall I wander?
> Upstairs and downstairs
> And in my lady's chamber

shows, in profile, steps being ascended on one side and descended on the other, the whole comprising a stepped hill, UPstairs, DOWNstairs. While reciting rhymes like this or "Wee Willie Winkie" (" . . . runs through the town, upstairs and downstairs, in his nightgown . . . ") the narrator's hands and voice move no differently than when conducting the little engine up over the mountain and down into the valley.

The point is not merely that the child is surrounded by legions of hills and mountains represented in profile, but also with . . . *a formula for drawing them.* The child has not only heard the unceasing "Uppp, dowwwn," but has heard, seen and felt the *concept* so encoded in (1) changes in vocal pitch, (2) illustrations, and (3) numerous kinds of gestures, not just in connection with hills or rhymes but in many situations. Whether from explanations of elevators or being tossed in the air, from picture books or explicit teaching about directions, UP and DOWN come to constitute a conceptual-gestural-vocal-graphic complex, which, given a pencil or crayon, just about automatically produces a simulacrum of a hill in profile. This is to say that a child's first pictures of hills might not be hills at all but records or traces of gestures—upp, dowwwn—which subsequently (or even during the act) *become discovered as hills.* The "tall hills" of several pages back could easily exemplify such discovered or recognized hills, as could the "mountains" that followed, clearly nothing but up-and-down variations in the forward motion of the line. That the syncreticism of these drawing is of this exclusively manual origin rather than pedal—as might be suggested from the direct acts of walking or being strolled up and down hills—is amply demonstrated by their profilism, which results *directly* from the suggested manual syncreticism, but would have to be a complex *derivative* of a pedal origin (involving rotations presumably beyond children of this age). Once recognized for what they "are," the ability to draw hills is a *fait accompli.* Exercise of the new-found skill will proceed to the elaboration of different kinds of hills, which, through the processes James called convergent evolution but Spinden involution,[75] will come increasingly to resemble—not the veritable hills of the child's underfoot experience—but the hills-in-profile of the picture books. Such are the mass of the hills we have seen.

The Hillsigning of the Contemporary Child in Context

Hillsigns are thus but a part of the larger system of representations—including words and numbers—that the child discovers and into which he is socialized. The question ceases being, "How does the child develop this image?" and becomes "Why is this the image we choose to teach him?" There can be no more question of a child *spontaneously developing* a hillsign than of *spontaneously developing* a word—such as "hill" for instance.[76] The reason the child's hillsign bears so little resemblance to his own experience is because, as with the word "hill," it was developed out of the *adult experience* of the entire culture. The child *learns* hillsigns, for like the greater language of which they're a part, they're not something that comes naturally to the child *qua* biologic entity, but something that comes laboriously to the child *qua* social person, for representation—whatever else it might be—is fundamentally a social fact, a *sine qua non* of communal existence. In a sense, then, except as their stages might indicate the sorts of cultural information capable of being *integrated* by the individual, structuralist models of cognitive development are somewhat beside the point. Despite their ability to ape observed sequences of change they are, in effect, too simple, consistently underestimating the significance of the child's . . . social situation, of the developing organism's . . . broad environmental context. Whereas there can be little doubt about the developmental *primacy* of the child's earliest hillsigns within a structuralist framework—the primitive syncreticism of the salient *up* and *down*, the egocentricity of its viewpoint, its concreteness and global quality are all characteristic (and after all, no one *taught* the first Mixtec, Mesopotamian or Egyptian hillsigners)—there can likewise be little doubt that the sign is manifested *in our sample* as a function of exposure to the sorts of kids' books denominated above; as a function, in general, of access to the cultural mainstream. Thus poorer children—with less exposure to children's books at home—lag three, four and even more years behind others in their ability to make even the most primitive hillsigns; and some high school students, having missed the earlier literature without having leaped to the more advanced forms of their quondam peers, are incapable of graphically representing landform relief at all. In adults this shows up in a complete refusal of the task—under any circumstances—in the well-documented: "I can't make a map. I can't draw. I've *never* been able to draw!"

Thus, whereas the profile hill will inevitably be the first hillsign to appear in a given developmental sequence—precisely because of its syncreticism, the egocentricity of its viewpoint, its concreteness, its globality—that it appears *at all* is a fact of social expediency and *not* developmental necessity: like multiplication, not like sex. Multiplication provides, in fact, an apt illustration. As is the case with the hillsign, multiplication "grows" in the child at a predictable point in a developmental sequence—between addition and long division—that happens to parallel the sequence observed in the history of arithmetic.[77] Like the hillsign, multiplication does not come naturally to the child, but laboriously. Most children must be forced to internalize the vast quantity of culturally developed information stored in the "times" tables, the possession of which, in Plato's day, would have rendered them prodigies of prodigies. In the case of multiplication it is obvious that, in Newton's phrase, we are standing on the shoulders of giants; but we are doing likewise in drawing profile views of hills at 3 and 4 years. Approaching our new query—why this is the hillsign we choose to teach our children first—from this direction brings us closer to an answer, for it is like asking why we choose to teach our children multiplication before long division, algebra or trigonometry.

The answer is embedded in both the structure of the information and the structure of its transmission—though these are scarcely independent. Multiplication, like the hillsign, does *seem* developmentally prior, simpler and more concrete. Long division subsumes multiplication, multiplication does not subsume long division. But these relations are not free of the effects of the sequence in which these operations were historically worked out. They are taught in this sequence, and they are related to each other as they are because this is the way they have always been related to each other, and this is the way they have always been taught. They are taught in this order because the teaching of long division has always taken multiplication as its *donnée*, and because multiplication has always had addition as its *donnée*.[78] And they are taught in this order because the most *broadly useful* social skills must be taught first in an educational system many enter but few leave: it is difficult to function in a monetary society without being able to add, hard to find a good-paying job without being able to multiply, difficult to get through graduate school without algebra and trigonometry, hard to work in the Pentagon without calculus. As A. N. Whitehead remarks, appositely enough for both hillsigns and arithmetic:

> By relieving the brain of all unnecessary work a good notation sets it free to concentrate on more advanced problems. . . . By the aid of symbolism, we can make transitions in reasoning almost mechanically by the eye, which otherwise would call into play the higher faculties of the brain. It is a profoundly erroneous truism, repeated by all copy-books and by eminent people when they are making speeches, that we should cultivate the habit of thinking what we are doing. The precise opposite is the case. Civilization advances by extending the number of important operations which we can perform without thinking about them.[79]

It is easy to create brilliant *ex post facto* explanations of this order solely in terms of the structural embedding of the information itself, but in the context of our historical situation—*in which these sequences are embedded in each of us through the structure of their transmission*—they are difficult to substantiate.[80] This difficulty, on the other hand, is not sufficient grounds for dismissal.

Since there is no reason to suppose that the logic of generation is anything like the logic of transmission, there is no reason to reject either of the explanations adumbrated, although there is plenty of reason to restrict their domains of influence. The structuralist argument seems strongest in explaining not ontogenetic but ethnogenetic development, whereas transmission logics (conservative, not generative) seem most successful in describing childhood acculturation (most of which is questionably mapped onto biologic maturation by the structuralists). However, it has to be admitted that insofar as the structuralist argument is used to describe the ethnogenesis of, say, hillsigns, *that it also is necessarily transmitted by the conservative mechanisms of acculturation*. Thus, not only are the *hillsigns* brought forward but also the *structurally determined order* in which they were originally elaborated. This simple embedding provides a complete explanation of the parallelism of ethno- and ontogenesis (of the history of hillsigning in the culture at large, and the history of its development in the individual), while further suggesting that whereas structuralist models might describe the genuinely *generative* behavior of individuals (most likely adults), it is conservative transmission models that are most powerful in describing the behavior of *acculturating* individuals (most typically children).[81]

The Development of Hillsigns

It is now possible not only to answer our two original questions—why the sequences are parallel and why they follow the order observed—but to provide, not just a sketch of the hillsigning complex, but a developmental model for cartographic signs in general. In brief, some 6,000 or 7,000 years ago in the Middle East (and at various times since in other parts of the globe), people in rapidly expanding and probably protourban societies experienced the compulsion to begin keeping records. Doubtless the reasons varied (but they were unlikely to be different from those that prompt us to keep records today). As we have seen, Schmandt-Besserat attributes it to the necessity for accounting in long-distance trade and Smith to the necessity for recording and legitimating ownership of land. In both cases, social and economic power was at stake. In each of these early notations, signs were originally as labile as a 3-year-old's image of a hill: a variety of modes ranging from the linguistic through the logographic to the purely pictorial—and including mixtures of each—must have been hazarded in the struggle to preserve both qualitative and quantitative information in both spatial (geographic) and temporal (historical) dimensions. Signs which sprouted as names (for example, in lineages) ripened as pictures (for example, on maps) . . . and *vice versa*. So the systems differentiated (as we can see them differentiate in children growing up today). Writing, with its logographic and *linguistic* means, grew into one immense branch; mapping and other spatialized arts with their logographic and *pictorial* means, into another, but both writing and mapping remained rooted in the same soil. One of these was the hillsign. Its pictorial character, there from the beginning in the Mixtec logogram or the Chinese ideogram, flourished in the microclimate of the map, spreading and evolving in its cartographic role as "map symbol."

This early hillsign took the form of a profile of a distant and highly generalized hill—susceptible of logographic and pictorial supplementation—*for precisely the reasons advanced by the structuralists*. Its form is gesturally syncretic with the act of manually tracing out a hill on the horizon as well as with the (rotated—no problem for an adult) act of mounting and descending even minimal relief. The most salient feature of a hill—its up-and-downness, its jutting out—is simply and directly presented in a form closely resembling representations of the nose and chin (also originally represented in profile).[82] Although modifiable through supplementation, the form is concrete and incapable of manifesting—in itself—variations in observed hill form. On the other hand, its general simplicity lends itself admirably to future development. It can easily be made larger or smaller. Is peak can be made sharper or rounder. Numbers of them can be strung together to represent a chain of hills, or with the advent of the use of the overlap cue, a range of hills receding from the viewer. Its form can also be changed to increasingly reflect observed profiles. Once the basic lump takes its secure place in the signage system, modifications in shape can move it farther and farther from its original form without incurring loss in recognizability. And yet, despite these developments, the sign retains its fundamentally concrete, gesturally syncretic and pictorial character.

Once developed, the sign, along with others in the broader synthesis of sign systems, would be passed on from generation to generation of priest, scribe or record keeper. It would not be taught to children. No matter how childish the sign might appear to us today, what child could have hoped to have penetrated the fathomless secrets—wrested with effort, insight and inspiration from nature, gods and the ages—of writing in any of its forms? One would as soon expect the operator of a funeral parlor to admit he's a tradesman—or a cartographer to admit he's a craftsman—as expect those early priest–scribes to admit the uninitiated to their sacred preserves.[83] But with the passage of time, the form would sufficiently evolve—and signage systems become sufficiently common—that this ancient form would acquire the stigma of "Well, that's how we *used* to do it," or "Here, *you* can play with this one." The newer form would draw on the inventions of general foreshortening, perspective systems and shading with their endless permutations and interpenetrations. Granted the earlier profile view, however, these new images present no radical challenge. *Within* the existent profile image one need merely shade, in even the most stereotyped and superficial manner, to provide what had been heretofore a croquet wicket (a one-dimensioned line) with a billowing tent-like cover (a two-dimensioned surface). The change reflects, of course, changes in visual perception and graphic representation (such as foreshortening), but it generally also

Although all these hills were drawn by tenth graders (ages 15–17) thirteen types of hillsigns are included in these eighteen hills covering all of the salient stages in the ontogenetic sequence.

reflects, in both onto- and ethnogenetic sequences, changes in the relationship with the landscape. A profiled hill hides *everything* behind it and reduces its surface to a single line: everything on the hill must appear on that *line* if at all. A hill with a surface falling toward the viewer has radically increased the area that can be occupied by things in their veridical locations, whereas its implications of obliquity result in the revelation of things that had been obscured *behind* the older hill. Maps employing the newer symbol inevitably portray *heightened environmental exploitation* of one kind or another, usually in response to "imperial" needs to assess taxes, wage war, facilitate communications or exploit strategic resources.[84] But before the new generations of scribes and mapmakers can learn the new oblique view, they have to be familiar with the older profile view. In learning as in discovering, edge precedes surface. The profile hill thus begins its descent down the pedagogic—and ontogenetic—ladders.

The implications of obliquity latent in the earliest shaded hills were lost on no one. Each of the advantages vouchsafed in foreshortening was enhanced as the bird's eye moved directly over the hill. From this position its entire surface could be veridically populated while the . . . other-side-of-the-hill vanished entirely. This new hill-in-plan could not, however, be represented by the same old extension of devices used to represent the hill in oblique. Although there can be little doubt that early attempts at hachures grew out of the hatching used as shading, the contour lines that ultimately supplanted them were borrowed from a development whose genesis is as unclear as its contributions are obvious (lability is the necessary character of innovation).[85] The two-dimensioned surface of the foreshortened view has turned into a three-dimensioned volume. But whatever the precise nature of the origin of the contour, instruction in its use is as completely dependent on a mastery of the oblique and shaded hillform as its was on a mastery of the profile. The distance separating this abstraction of relief from the profile view is frequently encountered on contemporary maps. Since a topographic survey sheet is not a picture of topography but a recording of arbitrarily selected lines of constant elevation, the relief itself—*qua* relief—must be reinterpreted from it. Thus, *as if models of hierarchic integration themselves*, topographic survey sheets frequently employ *vertical relief shading*—brought forward from the oblique phase and *repictorializing* the map—as well as *profiles* (dignified as cross-sections)—brought forward from the earliest phase—returning to the map the salient sense of UP and DOWN.

With the broadening of education in general and map use in particular, induction into the relevant signage systems has not just been pushed down the pedagogic and ontogenetic ladders—it has been pushed *off* the formal pedagogic ladder into the home. If typical 22-year olds are to be expected to use topographic survey sheets—as civil engineers, landscape architects, architects, planners, geographers, cartographers,

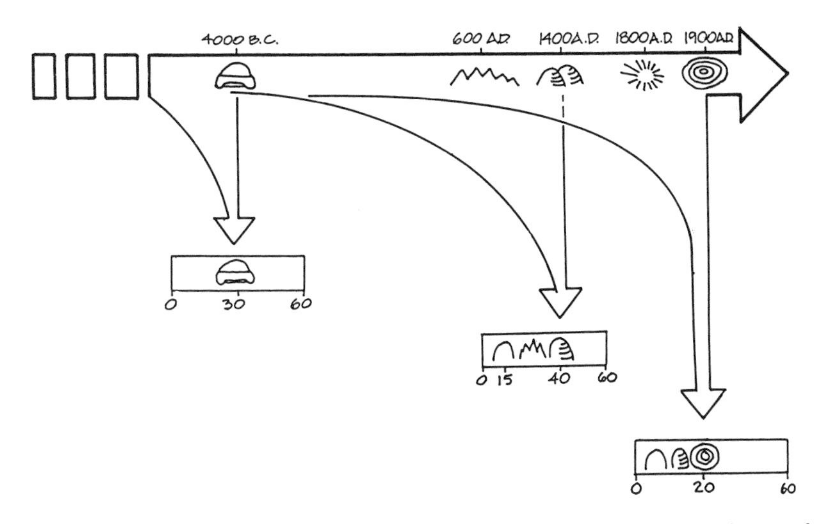

The upper arrow represents the ethnogenesis of hillsigns, with indicative dating of use only—not discovery. In lower oblongs, the historically occurring ontogenetic sequences are represented, with ages at which an individual might have mastered a hillsign complex noted.

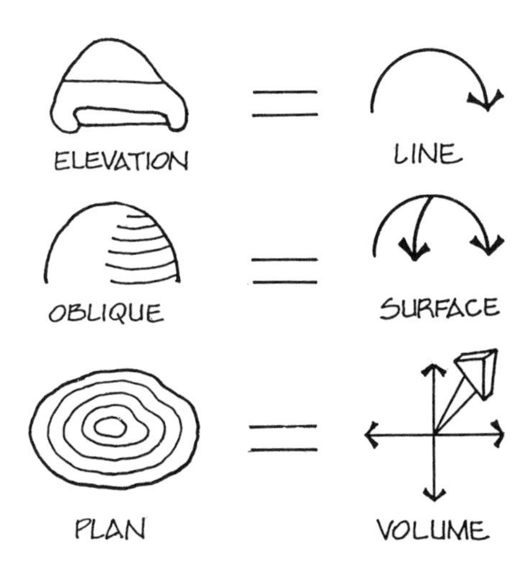

As the hillsigns develop on the left from elevation to plan, its ability to represent more fully the physical reality of our existence is increased. In elevation the hill is essentially the "line" of its silhouette; in oblique the hill is nothing but the "surface" of the presented slope; in plan—with contours—the hill can be seen as a "volume" for the first time. Future hillsigns mat opt to add something of geomorphic dynamism to the picture.

extension agents, *whatever*—introduction to the map signage system must begin early and informally, as early as children today are introduced to what was once the secret of secrets, the alphabet (or the number system). Parents, part of this larger system, orient their children to be effective learners. In selecting and performing *The Little Red Caboose*, parents lay the groundwork for—among endless other things—their children's ability to make, deploy and interpret the complete complex of hillsigns used by contemporary cartographers.

Hillsigns of the Future

The child is not, however—despite the size, power and anciency of the acculturating juggernaut—merely a passive recipient of the hill signage system. Although it is true that children as children do not extend and

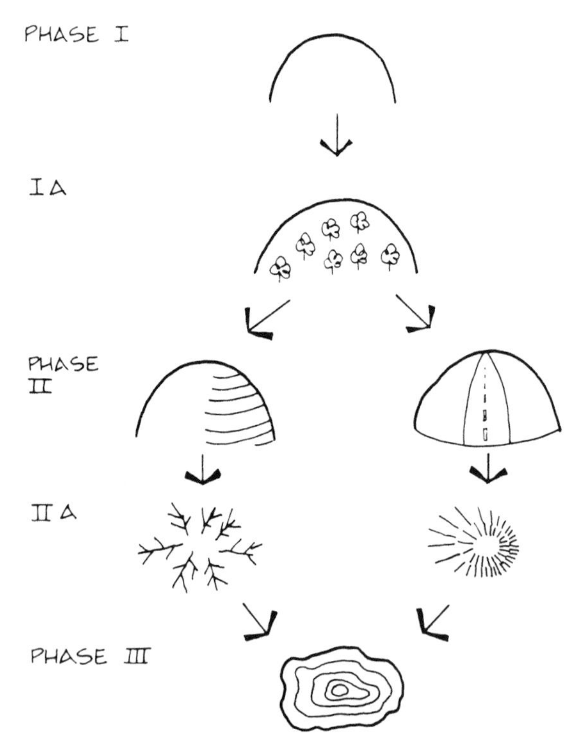

Ontogenetically, the hillsign unfolds in parallelism and mimicry the ethnogenetic sequence, with evident variations. The child's hill in Phase I is "curtained" in Phase IA before splitting in Phase II into a shaded hill(at left) and a foreshortened hill (at right). These develop in Phase IIA into either or both hachure signs illustrated before "maturing" into contours in Phase III.

enrich the system, they nevertheless differentially prepare themselves to do so by virtue of their relations to both hills and the act of signing. This is obviously true in a cross-cultural sense. The child of the Tzotzil Indian who says ". . . it is the mountain that protects life and also that the milpa is on the mountainside. It is there that our work is and consequently, our life . . ."[86] will relate to hills differently than someone raised on the shores of Lake Geneva or the west side of Cleveland. But more to the point is the fact that different kids on the west side of Cleveland relate to hills in different ways. Some exult in the long slide into the valley of the Cuyahoga—on sleds through the winter snows, on cardboard down the summer grasses—but some stand on the ridge in awful fear. Others avoid the slopes entirely. Not only do these sorts of variations condition the underfoot experiences that syncretically undergird the signs, but they also condition the effects made by the books and stories—a class variable in their own right—which in turn have their impact on the interpenetrations of hill symbolism with notions of religion (Calvary, the Ascension, the white church high on the hill), class (higher classes living higher up the hill) and simple struggle (after getting to the top you coast down the other side), as well as the more pedestrian issue of . . . the interpretation of contour lines. The point is that kids have *feelings* about hills that aren't simple and that are also part of the equation. Although it is probably worth repeating that in large part children are inducted into the signs they use—which provide the foundation for thinking about their own idiosyncratic experience—by a social system devoted to its own supraindividual and expedient ends, it is definitely worth illustrating something of the unravelable complexity that frequently works to save it from itself.

In the second growth woods not far from their suburban subdivision, five preteen boys have constructed a jump for their bikes by removing trees and grass and erecting posts, rails and signs. These are hill users. In their drawings and words you can see and hear it all, the deep enduring acculturation, the throbbing pulse of generation, the quiet voice of a genuine freedom:[87]

Q. What do you think this hill is made of?
A. Clay and dirt? Some other stuff . . . uh, dirt and rocks put together . . . Mud! Red clay, rocks, ummmmmmm parts of trees, limbs . . . probably small tires are lost down there, screw drivers . . .
Q. Well, where do you think it came from?
A. Well, the water ran out . . . washed the dirt away—Uh, I think it was probably a, uh, kind of like a plateau or something like that with trees and all on it—and some time ago they went through there with a bulldozer and cleaned it out down to the light down there—I know there used to be a house down there and somebody

used to, you know, they kept driving down there and I think they tore down the house.

Q. Have you changed the hill much?

A. Well, we pack the hills and then we clear out all the stuff and where we ride we move some of the big rocks—

Q. Pack?

A. Well, we get on our hands and sometimes we jump on them to make them, to make them pack where, where we won't slide or anything . . . And we put down straw to make it better to ride and . . . cut down trees and stuff . . .

Q. If you could have made the perfect hill, what would it have been like?

A. It would be, it would have flat land and then you'd have a hill about 6 feet high and about a 35° angle and hard dirt and uhhh, then at the top, just flatten out with no trees or rocks or stumps up there that you can wreck on . . . And have markers on the side where you could follow and know you were going . . . It would be really steep and packed good and about 10 feet high—yeah, about 10 feet high!

Q. Well, how do you feel when you go over the hill?

A. Well really sometimes my stomach gets to me and I hope I don't wreck, yeah, I get a nervous feeling in my stomach. You get kind

Part of a child's drawing, redrawn for reproduction's sake. Note the butterfly: it's actually a hillsign, not some local fauna.

of scared. You've got to have a little brave in you: once I was *dee-mol-ished* but I wasn't scared a bit, got right back up . . . Yeah, my stomach feels a little funny, like jiggling around inside of me—I don't know, it . . . when I start off . . . going down the hill . . . on a take-off it makes me feel kind of brave and all . . . and, uh, just before I hit the hill to jump, it makes me feel kind of scared I might wreck or something and tear up my bike or myself and then when I'm in the air, it makes, you know, it just feels like fun like . . . you're not, you're free and you know you go with the wind and all and when you touch ground safely . . . you stop . . . it makes you feel good that you jumped.

Q. When you are just going over the hill, what kind of feeling is that?

A. You get butterflies and all, in your stomach. . . .

It would be nice to imagine he felt those butterflies again when long years later, scrunched over a drafting table or propped before a computer monitor, he finally figured out how to put them on a map in the form of a hillsign. Yes, and perhaps he will. . . .

CHAPTER SEVEN

The Interest the Map Serves Can Be Yours

But it's too late for *us* to invent hillsigns like the boy. We're *way* too old. Long since have we prepared our face "to meet the faces that we meet" (we already wear the bottoms of our trousers rolled). We have heard the mermaids singing each to each: we know they will not sing to us. We know *too much*. The stink of interest lies in our nostrils. Not even the Surrealists could free the map of its reek. Look at theirs: how, in its espousal of their interest (the United States . . . *eliminated*, Easter Island exaggerated to the size of . . . *Europe!*), is it to be distinguished from the Robinson, the Mercator?[1] Even the Mixtec cartographer who 500 years ago made an earlier butterfly hill . . . *knew too much* (the butterfly hill was the name of a town, it was put on the map to mark its allegiance to Zacatepec, its subservience).[2] Interest pervades everything, it is everywhere, its weight is a burden, it oppresses.

So . . . maybe it is too late . . . *for that*. Then . . . why struggle to evade it? Why not admit the interest in the map, that very interest which by selecting and so making the map a *re-present*-ation *works* for . . . *Tom Van Sant*, for the . . . *National Geographic Society*, for the . . . *United States Geological Survey*, for the . . . *State of North Carolina*, for the . . .. *cacique of Zacatepec*. Once the map is accepted for the interested representation it is, once its historical contingency is fully acknowledged, it is no longer necessary to mask it. Freed from this burden of . . . *dissimulation* . . . the map will be able to assume its truest character, that of instrument for . . . *data processing*, that of instrument for . . . *reasoning about quantitative information*, that of instrument for . . . *persuasive argument*. Freed from the tyranny of the eye (the map never was a *vision*

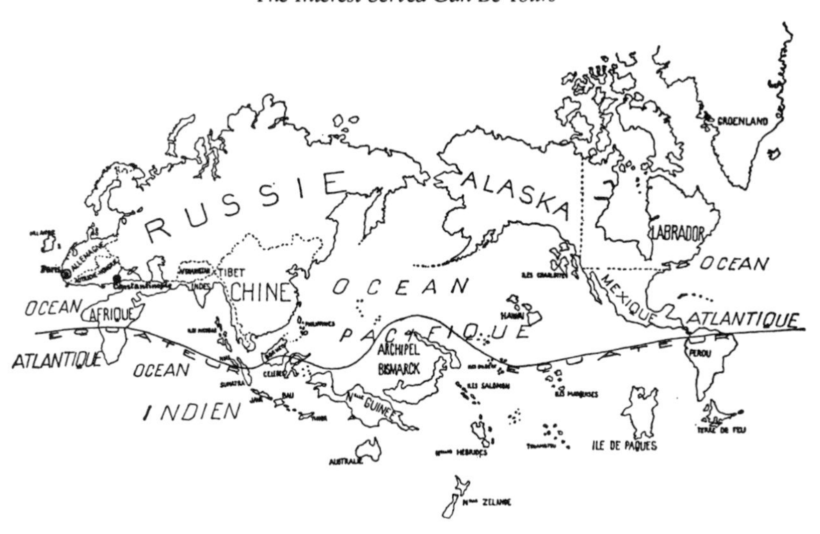

Surrealist map of the world. (From *Variétés*, Brussels, June 1929.)

of reality), the map can be returned to . . . *the hand* (that makes it) . . . *the mind* (that reasons with it) . . . *the mouth* (that speaks with it). Freed from a pretense of objectivity that reduced it to the passivity of observation, the map can be restored to the *instrumentality* of the body as a whole. Freed from being a thing to . . . *look at*, it can become something . . . *you make*. The map will be enabled to work . . . *for you*, *for us*.

This Mixtec sign for San Vicente Piñas consists of a hill and butterfly, expressing San Vicente's Mixtec name, *yucu ticuvua*, that is "hill of the butterfly". (Redrawn from Plate 109 of Mary Elizabeth Smith's *Picture Writing from Ancient Southern Mexico*, University of Oklahoma Press, Norman, 1973.)

Anybody Can Make a Map

Here, from this morning's paper, headlined, "Drug abuse rises in coastal areas: interstates still main corridors," an article that begins:

> With a map of North Carolina in one hand and a stack of charts in the other, Tony Mulvihill sat down over the weekend to hunt for patterns in illegal drug use across the state. He was startled by what he saw. Using a marker to shade in areas with the worst problems, Mulvihill colored large splotches around the state's big cities and along interstate highways. But to his amazement, he also found a huge dark swath running down the coast from the Albemarle Sound to the Cape Fear River.[3]

Here there is no question of . . . *looking something up*. You want to see what the pattern is, you take marker in hand, you color. Why? Because you want to shift resources, you want to persuade: "These are mostly very, very rural areas with few resources for dealing with the problem," said Mulvihill at a news conference. The point is publicity, at stake is money, the allocation of tax dollars. There is no question of detachment. This is an instance of advocacy. Is it true? *You* color the map, *you* call the press conference, *you press . . . your point of view*.

Does it have to be markers on shirt cardboard? No, it can be as slick as you like. The point is the acknowledgement of the interest (we want to see . . . *dueling maps*). Here, these are professionals (listen to the urgency in their voices):

> By the time this article appears, there will be approximately 172,000 people with AIDS in this country. How many more will eventually convert to AIDS from their infection with the human immunodeficiency virus (HIV) no one really knows . . .

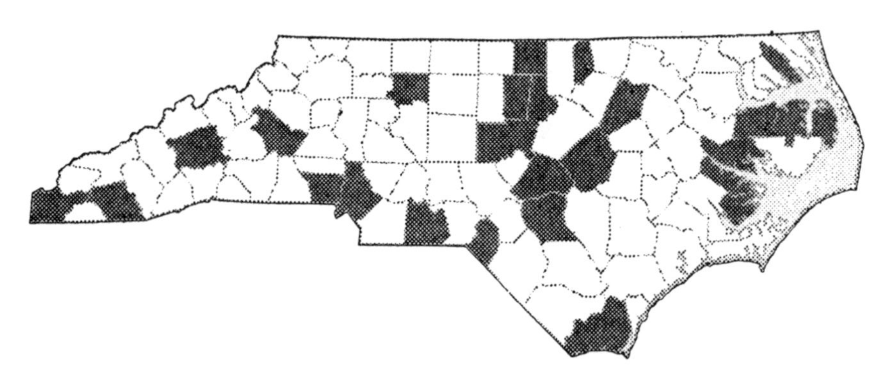

What started out with a handful of colored markers has ended up in this newspaper rendition of Tony Mulvihill's map as a much more authoritative image. (Courtesy *News and Observer*.)

> Why bother to predict the next maps of AIDS? For three good reasons. First, to achieve a deeper scientific understanding of the epidemic, we must know what is happening in space as well as in time. We have been astounded at the lack of geographic or spatial thinking in all the conventional approaches to modelling the epidemic over the past five years . . .
>
> Second, if truly effective educational intervention requires cues to action, then there is nothing like seeing is believing. Earlier we used the words *distant* and *remote* in referring to the way young people view the epidemic, and we used them with good reason. For many people, the epidemic appears far away, when in fact it is all around them. . . . When map sequences, including maps predicting for the future, are linked together in animated form and shown on television, you can tell from young people's reactions ("Wow, man, I never realized it was that close!") that you have caught their attention. They can see immediately with their own eyes the way in which the AIDS epidemic jumps from city to city in hierarchical diffusion, and then spreads out by spatially contagious diffusion from regional epicenters into the surrounding countryside, like a wine stain on a tablecloth . . .
>
> Third, most of the health care delivery systems in our major metropolitan areas are already under great stress from caring for people with AIDS, and with the sorts of monthly increases we are now seeing, the situation is going to get much worse. But planning for expanded facilities (hospital beds, hospices, and so forth) requires that we think about where we should put new facilities—the geographic question—not just when the expected numbers are going to appear. And our concern is not the usual one of economic efficiency, but for the totally humane reason of helping people—brothers, sisters, parents, wives, husbands, lovers—to have maximum access to those who are dying.[4]

There is no question here of markers on a map (but it could well have started that way). Peter Gould and his colleagues approached the problem by transforming geographical space, by equational expansion, by spatial adaptive filtering; and they produce their results in dramatic, colored animated sequences designed for television. In such hands, maps cease being ends in themselves.[5] They strive to become the "moment in the process of decision-making" that Jacques Bertin has insisted maps should be.[6]

Maps Are Moments in the Process of Decision-Making

Bertin has also insisted that maps are "not 'drawn' once and for all," but are "constructed and reconstructed until [they] reveal all the relationships constituted by the interplay of the data."[7] Since the map usually tries to pass itself off as a *picture* of the way things are, this is not

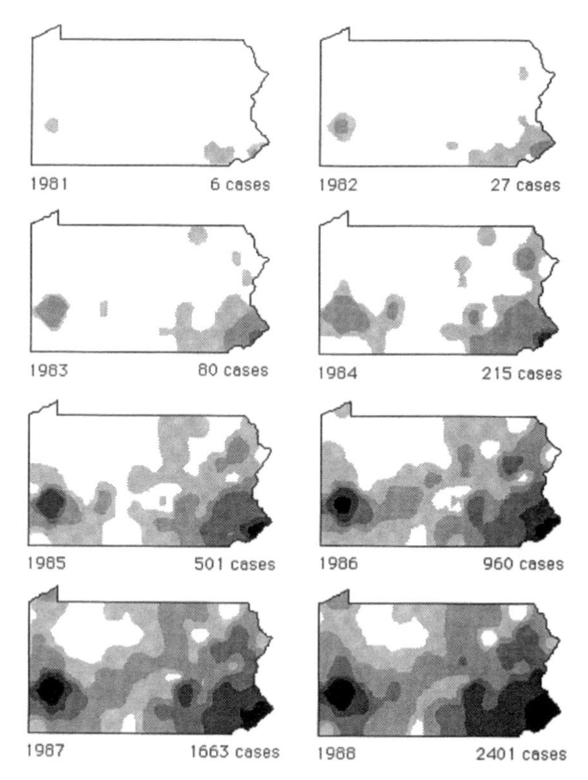

Diffusion of AIDS in Pennsylvania, taken from a colored and animated sequence made for television. Gray tones increase geometrically, 1, 3, 9, 27, 81, 243, and so forth. (Courtesy of Peter Gould.)

something you usually get to see; but when the issue is empowerment, the construction and reconstruction of the map is what it's all about. When the Detroit School Board's decentralization office adopted a redistricting plan required by Michigan law, Gwendolyn Warren and William Bunge not only came up with an alternative, they came up with lots of them, not just for the sake of having alternatives but because that's what you come up with when you try to understand all the relationships constituted by the interplay of the data. Here, let them describe it:

> In response to your request for technical assistance in the implementation of Senate Bill No. 635, we hand you herewith a copy of a report entitled "A Report to the Parents of Detroit on School Decentralization."
>
> The report is interesting in that it required some of the latest programming techniques in the most advanced languages available on the continent. Five or six university mathematical and geography

departments have worked on the high school and grade school based region problems. We would like to draw special attention to the work of Dr. John Shepard, the geographer from the London School of Economics who this year is fortunately on leave to Queen's University in Kingston, Ontario, and who threw himself and colleagues into the task literally night and day to meet the deadlines set by men of more practical day to day affairs.

Thank you for the opportunity to turn abstract science to good use.[8]

Among the 35 maps published ("Grade schools more than fifty percent filled with black children are shown in black," "Racial tension: each dot indicates an incidence of housing discrimination as reported to the Michigan Civil Rights Commission, 1968–1969," "Residences of school board members") were 14 redistricting plans as well as pages of computer printout of possible high school combinations ("Proposed Solution from the University of Washington," "Computer Evaluation of All Decentralization Possibilities"). No less than 7,367 maps satisfying the initial constraints were found (this is constructing and reconstructing the map!). Given these constraints, Warren and Bunge observed that:

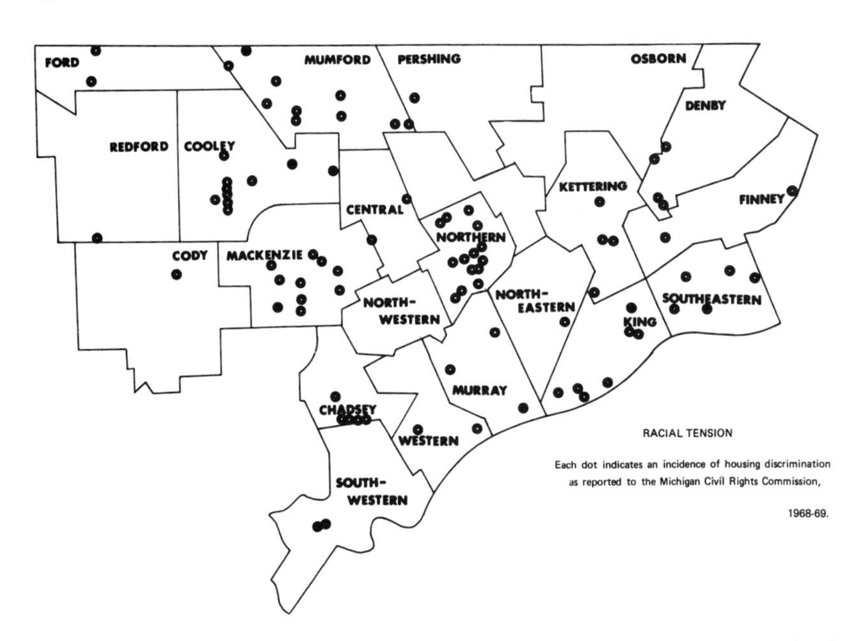

One of the maps from Gwendolyn Warren's and Bill Bunge's report on school decentralization, this one concerned with housing discrimination as an index of racial tension. Each dot indicates an incidence of housing discrimination as reported to the Michigan Civil Rights Commission, 1968–1969. (From the Detroit Geographical Expedition and Institute, *Field Notes: Discussion Paper No. 2, School Decentralization*, Detroit, 1970.)

> It is much easier to keep white children under white control than it is to protect black children from white racists. At the most, the black community can protect only 91.4% of its children, whereas the white community can retain control of 99.9% of theirs. At worst, the white community can lose control over only 45% of the white school children, although the black students can fall 75% under white control . . . Simply knowing how good or bad the final outcome can possibly be is a definite advantage in realistic discussions. We hope the city will utilize the research presented here to its fullest scientific extent.[9]

By developing—and exposing—the full range of solutions to the redistricting problem, Warren and Bunge pushed beyond advocacy into a kind of genuine professionalism, not the false kind consumed with techniques (the kind implied by the usual use of the term "professional cartographer"), but the kind implied by Bertin's dictum that, "A graphic is not only a drawing; it is a responsibility, sometimes a weighty one, in decision-making."[10]

Maps Are Heavy Responsibilities

But then this kind of responsibility has always been a hallmark of Bunge's work, and its implications for mapmaking have always been provocative. "What does a geographer mean by the statement that a portion of the earth's surface has been explored?" he has asked:

> Does he mean that the easy to map features for some harried early traveler such as rivers and mountains are accurately placed on a map? If so the earth is certainly explored. Humans are of great significance to geographers but are extremely difficult, even dangerous, to map. If the features of the earth's surface of interest to mankind include the human condition, then vast stretches of the map are in fact as "unexplored" as Antarctica in 1850 and should appear under that label and in the traditional intriguing chalk white color.[11]

Bunge recognizes that without a theme there is no map—the map is not of some*place* without being of some*thing*—and that the map is not somehow *innocent* with respect to this choice (it is not something *forced* on the map, the map comes into being in making it): "Geography is often defined as the study of the earth's surface as the home of man. But the view from which men's home? The perception from the homes of people that live in those particular places on the earth's surface, or rather from the homes of men in distant Buckingham Palaces or New York book publishers?" Nor, for Bunge, is it just a question of point of view. No aspect of the map is any more innocent. The map's scale, for example, has a way of determining—all by itself—what can be seen and what can't. For instance, at small scales kids just . . . *disappear.* They get swallowed up in the worlds of their parents. Therefore:

> Accusingly, there seems to be no geography of children, that is, the earth's surface as the home of children. What is their perception of their space? What is the "market area" of a tot lot? What is the average rate of travel of a kindergarten child? We seem to have ample statistics on the speed of trucks and giraffes. What is the traffic flow pattern of children across crowded streets including normally "illegal" children who jay walk and do other childish and disorderly things?[12]

Bunge's solution? To mount expeditions:

> To implement a truly human exploration of the earth's surface, the academic geographers, folk geographers, urban planners and others intrigued with such an effort, have founded the Society for Human Exploration. The functions of the Society are to assist exploration especially through the mounting of expeditions. The first of the planned series is the "Detroit Geographical Expedition, I" covering the entire urban conglomeration centered on Detroit. Its advance scouts are now in the field and completion date is projected for the fall of 1970.[13]

These expeditions did not denigrate local geographers to the rank of "native guides" and simply appropriate their maps of the Known World along with everything else. Rather the locals ran the expeditions, whose goal was the creation of *oughtness maps*. "After all," insisted Bunge, "it is not the function of geographers to merely map the earth, but to change it."[14]

Precisely the sentiment of Conservation International. They, however, are less interested in the earth's surface as the home of man than as the home of . . . *everything else*. Here, this is from their poster, *BioDiversity at Risk: A Preview of Conversation International's Atlas for the 1990s:*

> More species of plants and animals are in danger of extinction today than any time since the dinosaurs and their kin vanished 65 million years ago. Some biologists believe dozens of species are dying out each day—mostly in tropical rain forests. Since Conservation International was founded five years ago, we have been in the forefront of the fight to save these rich and endangered forests. But one organization, and indeed the entire conservation movement, can only do so much. Because time and resources are limited, we have had to make tough choices about where to focus our efforts.[15]

It's the same point made by Gould and his colleagues: effort in the wrong place is effort wasted. And again, economic efficiency is not the only issue: just as Gould wanted to maximize access to the dying for those who loved them, so Conservation International wants to minimize the destruction of areas of greatest biodiversity:

> In early 1990, for example, CI brought together a group of renowned field scientists to do fast but intensive surveys of unexplored tropical habitats as part of its Rapid Assessment Program (RAP). The first RAP team surveyed a 50,000-square-kilometer area in northwestern Bolivia near the Peruvian border. In this region, which had never before been considered for government protection, the researchers discovered one of the most biologically diverse rain forests on the continent. Their findings prompted officials to recommend protecting the area. Why had this biological goldmine been overlooked before? One of the most important reasons is that no botanists had ever worked in this part of the country, so the forest's extraordinary plant diversity was not known.[16]

What to do? One thing, obviously, is to get mapmaking technology into the hands of people who need it. RAP teams can assess an area's biodiversity, but then what? Conservation International

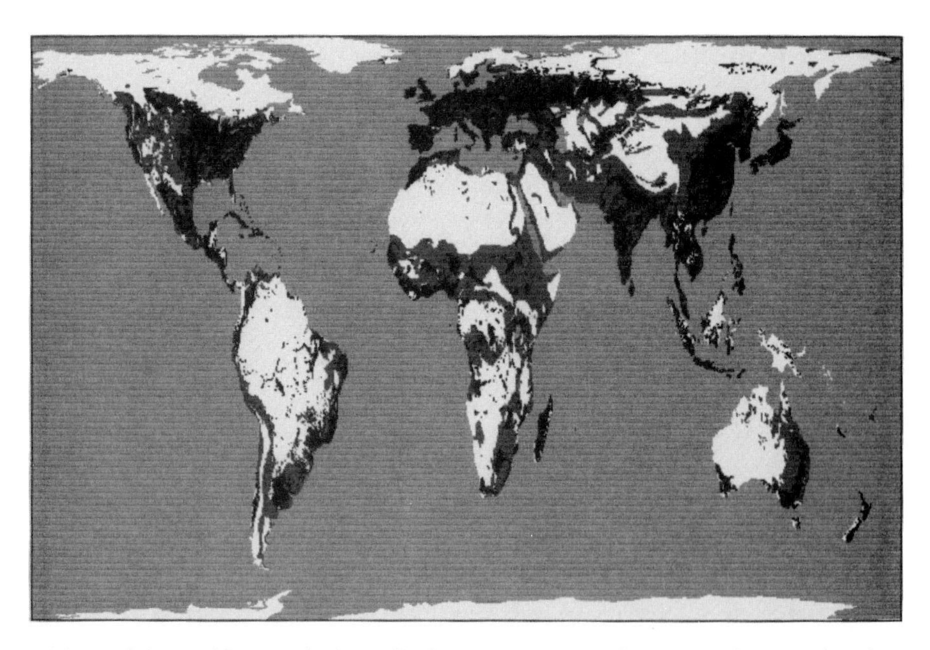

Many of the world's most biologically diverse regions are those most threatened with habitat destruction. Conservation International researchers took a significant step toward identifying habitats at greatest risk by charting data showing the distribution of human disturbance of ecosystems throughout the world. The darkest areas are human dominated, the areas in gray are partially disturbed, and the white regions remain predominantly natural.

distributes *CI-SIG*, a personal-computer based, multilingual mapmaking software package capable of integrating ecological data (conceivably developed by RAP teams) with social, economic and physical environment data. The result? Dramatic maps to help decision-makers better understand the problems of conservation and sustainable development. Nine of these maps are reproduced on this poster. Chromatically *HOT*, yet dressed in the "wet, ragged, long winter underwear" of the Peter's projection,[17] the poster combines the intellectual substance and the basement-press graphics of Bunge's *Nuclear War Atlas* (these on the poster self-evidently whispered their way out of a computer printer, you can see where the 8 1/2 x 11 sheets are pasted together) with the graphic sophistication of Herbert Bayer's *World Geo-Graphic Atlas* (the poster was designed by Kinetek Communication Graphics):

> This poster—the first phase of a new atlas project—illustrates some of the most important criteria CI uses to set conservation priorities. Relying on the most recent data on biological diversity, human disturbance of ecosystems, and in-country conservation capacity, our scientists have mapped the worldwide distribution of these important indicators using a Geographic Information System computer program. Together, the maps on this poster tell a grim story . . .[18]

In this program maps and mapmaking are everywhere: in the generation of the data, in the allocation of resources, in the publishing to raise awareness and money of this poster that promotes mapmaking as a way to stem the threats to biodiversity (the program is self-conscious about its use of maps). The interest, the instrumentality is not just palpable, *it is out front, it is part of the program*, not in the vaguely (or not so vaguely) self-congratulatory way of Tom Van Sant's World GeoSphere project (which in the end is . . . *art*), but in a simple pragmatic way, like Bunge with his expeditions . . . *getting the tools into the hands of those who need them* (and what is Conservation International's effort but a whole series of Bungian expeditions?), but more sophisticated tools, like those Gould and his colleagues are using to map the spread of AIDS. There is nothing reticent about *any* of this. Contrast it to the usual map or atlas that refuses to acknowledge its sources (the poster has a 37-item bibliography, it displays the linear regression equation for the species-area relationship), much less . . . *unmask its interest*.

Here: in contrast, a larger map, really quite beautiful, of Sikkim, published by the Association of American Geographers, with an "explanation" (instead of a "legend"), with all kinds of neat things on it, "ropeway", for instance, and "Levels of Settlement Hierarchy" (instead of "Cities"), with "Smaller Centers (Class 'C' Bazaar)" noted, and . . . well, anyhow, here's how it acknowledges its sources: "Compiled from best available maps and other sources supplied by the Government of

Sikkim." *Best available maps:* really, it's . . . *too much*. "Supplementary information on transportation network and settlement hierarchy were collected during field studies in Sikkim during 1965 and 1968. Base outline has been prepared from the Survey of India maps." That's . . . *it*. Touching detail: next to the index map showing the location of Sikkim, a . . . *reliability diagram*. The upper portion is pink ("Fair"), the lower part yellow ("Good"). What is this? A comment on the author's field work? An analysis of the "best available maps"? A passing reference to the quality of the Survey of India? Who knows? *Who cares?* It's just window dressing, a stamp of scientificity, *it means nothing*. As for the map's *reason for being* . . . in the acknowledgments, this note: "The cartographic work was financed from the University of Kentucky research funds." Why? Again . . . *who knows?* Whatever the purpose may have been, it is not acknowledged on the map. All the map acknowledges, *what it rubs in our face*, is the skill it took to make it, secret because invisible. In their impeccable shaded relief, the Himalayas all but burst through the paper. The cartographer has vanished (the more effectively to haunt our dreams); all we can see is . . . Sikkim.[19]

Maps Empower . . . By Working

Now contrast this again with Conservation International's poster: "All you need to make maps like *these*," it says, " is this software package we want to supply you with." Evaporated is the question of the cartographer's patient skill, his professional secrets, the hermeticism of graticule and graduated circles. Of course this terrifies the professionals (what will they do for a living?). Mark Monmonier anxiously wrings his hands over the inevitable "self-deception":

> Moreover, because of powerful personal computers and "user-friendly" mapping software, map authorship is perhaps too easy, and unintentional cartographic self-deception is inevitable. How many software users know that using area-shading symbols with magnitude data produces misleading maps? How many of these instant mapmakers are aware that size differences among areal units such as counties and census tracts can radically distort map comparisons?[20]

Better to trust your expert cartographer—with his "best available" data, his diminished Africa and his exaggerated Russia, his cloudless skies, his maps of potential . . . *iron mines*—than hazard an . . . unintentional self-deception. What is this? It is the heart of the darkness of our times, the assurance (and arrogance) of the . . . *expert* . . . that he knows . . . *better than you:*

> The professionals, that is the skilled and learned experts who apply their knowledge to the affairs and in the service of others, are traditionally held in high esteem. For generations, divinity, the law, medicine and even the military and now the newer professions in the fields of education, welfare, architecture, industrial management etc. have been acknowledged as being selflessly devoted to the good of the weaker and less knowledgeable members of society, thus enabling those who lack the capacity to fend for themselves to lead fuller, safer and healthier lives. However, the question must now be asked whether the professions in fact provide their services so altruistically, and whether we are really enriched and not just subordinated by their activities. There is a growing awareness that during the past twenty years or so, the professions have gained a supreme ascendency over our social aspirations and behaviour by tightly organizing and institutionalizing themselves. At the same time we have become a virtually passive clientèle: dependent, cajoled and harassed, economically deprived and physically and mentally damaged by the very agents whose raison d'être it is to help us.[21]

And who fits this description better than the cartographer?

But it's a ruse. We don't have problems cartographers can fix, we have *problems*—for example . . . *illegal drug use* (or AIDS or school integration or local self-knowledge or declining biodiversity)—and maps can help solve them. And it doesn't matter whether we make these maps with a marker or a personal computer, with a RAP team or an expedition into the heart of the cities we "live" in:

> Maps, by and large, are designed to show strangers the way. Now, the New Thing Art and Architecture Center has come up with a map of the Adams-Morgan community drawn for the people who live there. "To begin with," said Topper Carew, director of the New Thing, "we wanted to give our community a picture of itself—to define our territory."[22]

John Wiebenson, an architect, drew the map, and he did it with his hands and mind, his mouth and his feet:

> When "Wieb" (as he's called) wasn't talking, he was walking. "The best way to learn about any neighborhood is the way small boys do it—explore," he said. "I found all sorts of neat things that only the children know about—such as the steps that go down the steep slope from Waterside Drive to Rock Creek Park. They're great grandstand seats to watch the subway construction."[23]

All this walking and talking led to the discovery and solution of problems other than that of self-identity (it led to the design of a street signage system, proposals for a scenic path and viewing platforms), but then . . .

mapping doesn't just produce maps, it is fundamental to the process of bringing order to the world. Ten thousand copies of the map were mailed to neighborhood residents and the whole thing, mapping, mapmaking, printing and mailing came to less than $5000.

Nor do the maps we need have to be of our neighborhood, or even of what we usually think of as space:

> The space we live in is not always visible. More and more often, events that affect our lives in tangible ways occur in the invisible part of the geography. When you pick up the telephone, graze through endless channels of satellite or cable TV, buy something with a credit card, you are travelling in digital communication-space. It's time for people to find ways to visualize and map the new geographies of telecommunication. The authors have taken a first step by mapping the worldwide flow of telecommunications. In fact, a telegeographical overlay seems essential to any planet-monitoring information system, for telegeography is a concrete way of looking at the otherwise abstract territory where global-level self-awareness could take place.[24]

The accompanying map—created by a lawyer and a journalist and worlds apart from the cute architect's view of Adams-Morgan—shows, on a projection centered on the pole, the flows of phone calls coursing between the United States and Europe (on the one hand) and the Pacific rim (on the other). Rarely has the United States seemed so much . . . *in the center of things*, even without occupying that position graphically. The global electronic *village* that Marshall McLuhan promised is hard to see here. What shows up is the same tired hierarchy, only more dramatically, more pointedly than ever. "The global telecom network is really a network of networks," the mapmakers write. "It comprises over 540 million telephone exchange lines linking over 1 billion terminals in 200 countries. Some groups control the networks, while others are on the periphery."[25] And they urge that:

> Despite our daily experience of "globalization" and the "information revolution," many people still find it difficult to picture what is happening; they lack the right mental maps. Most atlases are of little help: for much of its history, geography has been preoccupied with political boundaries (which slabs of territory are controlled by which nations) and natural ones . . . By mapping network boundaries rather than political or physical ones, telegeography can capture the underlying flows which drive contemporary economic and political events. Moreover, by focusing on the pattern of networked communications—on what is connected rather than what happens to be physically conterminous—telegeography can help people to navigate this novel terrain.[26]

Novel terrain: but . . . *isn't it always?* This telecom network is not the first to connect the world. Bunge saw years ago that their missiles made the *surfaces* of the United States and the Soviet Union conterminous (Bunge's interest in making this map was . . . peace).[27] As long ago as 1864, Charles Minard drew a flow map of the export of French wines (Minard's interest was . . . statistical graphics).[28] Edmund Halley's 1686 map of the trade winds is an even more complicated map of connections, implying in those of the atmosphere those of the ships the atmosphere moved (Halley's interest was in . . . natural philosophy).[29] What is Matthew Paris's 13th-century itinerary but the map of a connection (Paris's interest was in getting from London to Rome)?[30] What else is the Roman Peutinger table with its road system tying together the Empire (this map's interest was in potential movement)?[31] Go back as far as you choose: what else has the map ever done but link—*connect*—the territory with what comes with it . . . with phones, if that's what it is, but the spirit world, if *that's* what it is. In no walk of life have people failed to use the power of the map to connect themselves to the world. Lawyers and journalists do it, conservation ecologists and architects do it, community activists and geographers do it. Yes, *even cartographers*, do it.

A legion of mapmakers, bewildering in their variety: *this* is the world of maps.

Notes

Chapter One

1. *Life, 15*(4), April, 1992, pp. 30–37.

2. Arthur Miller, *Timebends*, Grove Press, New York, p. 594.

3. "Global problem: Young students think earth is flat," *Raleigh Times*, March 7, 1988, p. 2. Lightman and Sadler originally published the results of their research in *Science and Children*.

4. Stephen Hall, *Mapping the Next Millennium*, Random House, New York, 1992.

5. Michael Kidron and Ronald Segal, *The New State of the World Atlas*, revised and updated, Simon and Schuster, New York, 1987.

6. Here's what the editors of *The Times Atlas of the World, Seventh Comprehensive Edition* (Times Books, London, 1985) have to say about the issue: "In recent years much political significance has been attached to the manner in which international boundaries are depicted and the way names are spelled in atlases. The position of *The Times* as publishers of this and all other atlases has been stated repeatedly and unequivocally. To attempt to judge the rights and wrongs of territorial disputes is beyond the function of the publishers of an atlas . . . In its atlases *The Times* aims to show the territorial situation obtaining at the time of publication without regard to the *de jure* situation in contentious areas or rival claims of contending parties." But then how do they decide where to put the border?

7. Stephen Langdon, "An Ancient Babylonian Map," *Museum Journal, 7* (1916), pp. 263–68. See also Jacob Finkelstein, "Mesopotamia," *Journal of Near Eastern Studies, 21* (1962), pp. 73–92, and A. R. Millard, "Cartography in the Ancient Near East," J.B. Harley and David Woodward, editors, *The History of Cartography Volume One: Cartography in Prehistoric, Ancient and Medieval Europe and the Mediterranean*, University of Chicago Press, Chicago, 1987.

8. For a handy summary see the entry in Helen Wallis and Arthur Robinson, editors, *Cartographical Innovations: An International Handbook of Mapping Terms to 1900*, Map Collectors Publications, in association with the International Cartographic Association, Tring, 1987. For the ancient near east, see O. A. W. Dilke, *Greek and Roman Maps*, Cornell University Press, Ithaca, 1985; for Japan, see M. Ramming, "The Evolution of Cartography in Japan," *Imago Mundi, 2* (1937), pp. 17–21; for Mexico, see Howard Cline, "The

Oztoticpac Land Map of Texcoco, 1540," in Walter Ristow, editor, *A la Carte: Selected Papers on Maps and Atlases*, Library of Congress, Washington, 1972.

9. Edward Espenshade and Joel Morrison, editors, *Rand McNally Goode's World Atlas, 16th Edition*, Rand McNally, Chicago, 1982, pp. 2–52.

10. Michael Kidron and Dan Smith, *The New State of War and Peace: A Full Color Survey of Arsenals, Armies, and Alliances Throughout the World*, Simon and Schuster, New York, 1991.

11. William Bunge, *Nuclear War Atlas*,Basil Blackwell, Oxford, 1988. This is an expanded version of Bunge's original 1982 poster version with its 28 maps.

12. James Scovel *et al.*, *Atlas of Landforms*, John Wiley & Sons, New York, 1965.

13. Hugh Johnson, *The World Atlas of Wine*, Simon and Schuster, New York, 1971.

14. Espenshade, *op. cit.*, p. 72.

15. Fritz Muhlenweg, *Big Tiger and Christian*, Pantheon Books, New York, 1952, p. 32.

16. *Ibid.*, p. 218.

17. "Gulf Islands: Victoria Harbour to Nanaimo Harbour," *Strait of Georgia Small-craft Chart*, Canadian Hydrographic Service, Department of Fisheries and Oceans, Ottawa, 1979, 1985 reprint edition.

18. These are no more than the marks included in the legend on the chart itself. A full list is available as *Chart No. 1: Symbols and Abbreviations Used on Canadian Nautical Charts*, Canadian Hydrographic Service, Minister of Fisheries and Oceans, Ottawa, 1984.

19. William Wilson and P. Albert Carpenter, *Region J Geology: A Guide for North Carolina Mineral Resource Development and Land Use Planning*, North Carolina Geological Survey Section, North Carolina Department of Natural Resources and Community Development, 1975, revised, 1981. The included map carries the additional authorship of John Parker.

20. The state of the art here is Tony Campbell's "Portolan Charts from the Late Thirteenth Century to 1500," in Harley and Woodward, *op. cit.*, pp. 371–463.

21. This is the standard Piagetian sequence. See Jean Piaget and Bärbel Inhelder, *The Child's Conception of Space*, Routledge and Kegan Paul, London, 1956.

22. For a general introduction to mental maps, see Roger Downs and David Stea, *Maps in Minds: Reflections on Cognitive Mapping*, Harper and Row, New York, 1977. I used the image of the Paris metro board in Denis Wood, *I Don't Want To, But I Will*, Clark University Cartographic Laboratory, Clark University, Worcester, Massachusetts, 1973, p. 22. For a more recent treatment see Denis Wood and Robert Beck, "Janine Eber Maps London: Individual Dimensions of Cognitive Imagery," *Journal of Environmental Psychology*, 9 (1989), pp. 1–26.

23. Joel Makower, editor, *The Map Catalogue: Every Kind of Map and Chart on Earth and Even Some Above It*, Vintage Books, New York, 1986.

24. William Loy, editor, "U.S. National Report to ICA, 1987," *The American Cartographer*, *14*(3), July 1987.

25. Arthur Robinson *et al.*, *Elements of Cartography, Fifth Edition*, John Wiley, New York, 1984. I quote from the section, "Classes of Maps," pp. 6–11.

26. Hall, *op. cit.*, pp xv–xvi.

27. Espenshade and Morrison, *op. cit.*, pp. iii–v.

28. Michael and Susan Southworth, *Maps: A Visual Survey and Design Guide*, Little, Brown, Boston, 1982.

29. J. B. Harley, "Text and Contexts in the Interpretation of Early Maps," in David Buisseret, editor, *From Sea Charts to Satellite Images: Interpreting North American History Through Maps*, University of Chicago Press, Chicago, 1990, pp. 3–4.

30. In Chapter Six the very hills will be acknowledged as social constructs.

31. The map not only creates the boundaries, but through them the thing distinguished: "Long before politicians created the Adirondack Park by encircling the region with the Blue Line, geological forces had formed the boundaries that define it," is the off-hand way Yngvar W. Isachsen puts it, failing to mention only the map on which the Blue Line was drawn. (Yngvar W. Isachsen, "Still Rising After All These Years," *Natural History*, May 1992, p. 31.) More generally, Brian Harley asks, "Could it be that what cartographers do, albeit unwittingly, is to transform by mapping the subject they seek to mirror so as to create not an image of reality, but a simulacrum that redescribes the world?" (J. B. Harley, "Can There Be a Cartographic Ethics?" *Cartographic Perspectives, 10*, Summer, 1991, p. 13), though there's nothing "unwittingly" about it.

32. This literature is vast. Classical texts include E.H. Gombrich, *Art and Illusion: A Study in the Psychology of Pictorial Representation*, Pantheon, New York, 1960; R.L. Gregory, *The Intelligent Eye*, McGraw-Hill, New York, 1970; Roland Barthes, *Mythologies*, Hill and Wang, New York, 1972; John Berger, *Ways of Seeing*, Viking, New York, 1973; Michael Foucault, *The Archeology of Knowledge*, Pantheon, New York, 1972. No one school, no one discipline: pervasive perspective.

33. Mark Monmonier, *How To Lie With Maps*, University of Chicago Press, Chicago, 1991. To his credit, Monmonier deals with more than the straightforward propaganda map, but the title of his book gives his thesis away: lying with maps is something you have to strain to do, not something that happens . . . *as a matter of course*.

34. For example, John Noble Wilford, caught in a deep rut, characterizes the Middle Ages as a "thousand-year slough of intellectual stagnations" and as "a millennium without a significant advance in the mapping of the world," in his *The Mapmakers*, Knopf, New York, 1981, p. 34. P. D. A. Harvey, David Woodward and Tony Campbell set him straight in their splendid contributions to Harley and Woodward, *op. cit.*, pp. 283–501.

35. The normative history of cartography is a ceaseless massaging of this theme of noble progress. "The history of cartography," writes Gerald Crone, "is largely that of the increase in the accuracy with which . . . elements of distance and direction are determined and . . . the comprehensiveness of the map content." (*Maps and Their Makers*, Dawson, Folkestone, Kent, 1953, p. xi). Lloyd Brown's *The Story of Maps* (Little Brown, Boston, 1949) is the classical example, to which Wilford (*The Mapmakers, op. cit.*) and even Hall (*Mapping the Next Millennium, op. cit*) self-consciously refer. It is *the* dominant theme.

36. *C&GS Update*, 4(2), Spring, 1992, p. 2. Contemporary degrees of precision are genuinely awesome. They also constitute an almost masturbatory precisionism for precision's sake.

37. Another example from the morning paper: the Raleigh suburb of Cary is considering commissioning a special census because it doubts the last U.S. Census got it right. Debbi Sykes, "Cary willing to wager U.S. Census is wrong," *News and Observer*, April 20, 1992, p. 3B. How wrong? Dead wrong. In Grand Rapids, Michigan, the Census Bureau reported seven people living in a cemetery: "Someone could have transposed a number," a spokesperson said. "Maybe it was a worker who was in a hurry to get home before the baby sitter left. It had the effect of moving those seven people from one block to another." Still, we sure know where that block is. ("Census dead wrong in Michigan," *News and Observer*, April, 22, 1992, p. 7A.

38. Though this a point we will stress only later, accuracy itself is of course a social construct. See Donald MacKenzie, *Inventing Accuracy: A Historical Sociology of Nuclear Missile Guidance* (MIT, Cambridge, 1990).

39. *Op. cit.*, Plate 86.

40. Debbi Sykes, "Raleigh neighbors don't want place on city's map," *The News and Observer*, April 18, 1992, p. 1B.

41. For example, Robinson *et al.*, *op. cit.*, pp. 7–9; Erwin Raisz, *Principles of Cartography*, McGraw-Hill, New York, 1962, p. 9. (Incidentally, Raisz has another tripartite "classification of maps": 1. General maps; 2. Special maps; 3. Globes and models.)

42. For example, in *Principles of Thematic Map Design* (Addison-Wesley, Reading, Massachusetts, 1985), Borden Dent opens his treatment of the thematic map with a discussion of "kinds of maps." This opens with two paragraphs on "general-purpose maps" so that the subject of the balance of the book can be opened with the phrase, "The other major class of map . . ." (pp. 5–6).

43. I refer here to the maps that embody the fictional worlds of Oz, Middle-earth, Lemuria and the like. See J.B. Post, *Atlas of Fantasy*, Mirage Press, Baltimore, 1973; its revision, Ballantine Books, New York, 1979; and Alberto Manguel and Gianni Guadalupi, *The Dictionary of Imaginary Places*, Macmillan, New York, 1980.

44. There is no question that this *author* increasingly resembles at best the director of a movie, at worst the entire crew, as responsibility for the map is diffused through its increasing intertextuality and among its increasingly specialized labor producers.

45. As we develop our analysis we will be able to get more specific about these. In Chapter Five we will attempt to demonstrate that the subject, for example, arises out of the operation of *four* codes, two of which, the *tectonic* (concerned with the space of the map) and the *temporal* (concerned with time) work at the level of language, as it were within the map (we call these codes of *intrasignification*); and two of which, the *topic* (the code of places) and the *historic* (the code of history), exploiting these, cantilevered off of them, work at the level of *myth*, so to speak . . . outside the map (we call these codes of *extrasignification*). Similar analyses are performed with respect to the theme and the author.

46. Espenshade and Morrison, *op. cit.*, p. vii.

47. *Ibid.*, p. 73.

48. Theodore Roszak, *Where the Wasteland Ends: Politics and Transcendence in Postindustrial Society*, Doubleday, New York, 1972, p. 408.

49. Donald Westlake, *High Adventure*, The Mysterious Press, New York, 1985, pp. 139–40.

Chapter Two

1. Tom Watterson, *The Revenge of the Baby-Sat*, Andrews and McMeel, Kansas City, 1991, p. 86. A few years ago Iona Opie made an identical point in a *New Yorker* interview: "I have a title for the new book. May I tell it to you? It is 'The People in the Playground.' 'People,' you know, because children *never* refer to themselves as children. They say, 'You need four people for this game,' or 'Some people say that rhyme differently'" ("Playground Person," *The New Yorker*, November 7, 1988, p. 31).

2. Appropriate here is A.A. Milne's reminder in "The End": "When I was One,/I was just begun./When I was Two,/I was nearly new./When I was three,/I was hardly Me./When I was Four,/I was not much more./When I was Five,/I was just alive./But now I am six, I'm as clever as clever./So I think I'll be six now for ever and ever." But the present always feels like this. "The End" concludes Milne's *Now We Are Six* (E.P. Dutton, New York, 1927, p.102).

3. The issue of scale in science is treated by John Tyler Bonner in his *The Scale of Nature*, Harper and Row, New York, 1969. In the domain of biology alone, the issue of scale is well treated in Thomas McMahon and John Tyler Bonner's *On Size and Life*, Scientific American Books, New York, 1983.

4. For a general statement of this theme see Ilya Prigogine, *From Being to Becoming: Time and Complexity in the Physical Sciences*, W. H. Freeman, San Francisco, 1980.

5. The popular introduction is Stephen Hawking's *A Brief History of Time: From the Big Bang to Black Holes*, Bantam, New York, 1988.

6. See Ronald F. Fox, *Energy and the Evolution of Life*, W. H. Freeman, San Francisco, 1988; but also Ernst Mayr's *The Growth of Biological Thought*, Harvard University Press, Cambridge, 1982.

7. See Humberto Maturana and Francisco Varela's brilliant *The Tree of Knowledge: The Biological Roots of Human Understanding* (New Science Library, Boston, 1988) for an argument that encompasses all these levels.

8. Christopher Tolkien, *The History of Middle Earth, Volume VII: The Treason of Isengard*, Houghton Mifflin, Boston, 1989, p. 295.

9. For insurance maps, see the entry in Helen Wallis and Arthur Robinson, editors, *Cartographical Innovations*, Map Collector Publications, in association with the International Cartographic Association, Tring, 1987, pp. 109–111; as well as, Robert Karrow and Ronald E. Grim, "Two Examples of Thematic Maps: Civil War and Fire Insurance Maps," in David Buisseret, editor, *From Sea Charts to Satellite Images: Interpreting North American History Through Maps*, University of Chicago Press, Chicago, 1990.

10. The classic texts here are Thomas Gladwin's *East Is a Big Bird* (Harvard University Press, Cambridge, 1970) and David Lewis' *We the Navigators* (University of Hawaii Press, Honolulu, 1972). See also the chapter on human navigation in Talbot Waterman's *Animal Navigation* (Scientific American Library, New York, 1989).

11. See, *inter alia*, Robert Rundstrom, "A Cultural Interpretation of Inuit Map Accuracy," *Geographical Review*, 80(2), April 1990, p. 157 and 163–166.

12. Robert Beck and I have closely analyzed the way sketch maps grow, in the case of one individual studying several of her maps. See Denis Wood and Robert Beck,"Janine Eber Maps London," *Journal of Environmental Psychology*, 9, 1–26.

13. Robert Rundstrom, *op. cit.*, 155.

14. The claim is made explicitly and at length in James Gould and Carol Gould, *The Honey Bee*, Scientific American Library, New York, 1988. See also F. C. Dyer and J. L. Gould, "Honey Bee Navigation," *American Scientist, 71*, 1983, pp. 587–597.

15. David Stea, "The Measurement of Mental Maps: an Experimental Model for Studying Conceptual Spaces," in K. R. Cox and R. G. Golledge, eds., *Behavioral Problems in Geography: A Symposium, Studies in Geography #17*, Northwestern University, Evanston, 1969, p. 229.

16. Waterman, *op. cit.*, p. 178. Waterman is explicit about the connection, noting that psychologists refer to the sort of "remembered spatial relations" he is concerned with as "cognitive maps." He does this on p. 182.

17. The literature is extensive. In addition to that cited by Waterman, *op. cit.*, and Gould and Gould, *op. cit.*, see D. R. Griffin, *Animal Thinking*, Harvard University Press, Cambridge, 1985; and J. T. Bonner, *The Evolution of Culture in Animals*,Princeton University Press, Princeton, 1980. More technical but no less useful are E. W. Menzel's "Cognitive Mapping in Chimpanzees," in S. H. Hulse *et al.*, eds., *Cognitive Process in Animal Behavior*, Lawrence Erlbaum, Hillsdale, N. J., 1978, pp. 375–422, and M. Konishi's "Centrally Synthesized Maps of Sensory Space," in *Trends in Neuroscience*, 9(4), 1986, pp. 163–168. There has been an explosion of work implicating the hippocampus (part of the posterior forebrain) in mammal orientation, taking off from John O'Keefe and Lynn Nadel's provocatively titled "Maps in the Brain," *New Scientist*, 27 June, 1974, pp. 749–751 (but watch for the dual use of "map" in this literature). Are animal maps linear or areal? Waterman says, "No doubt animals have both linear and two dimensional types" (p. 179), where what strikes me is that this question, so often raised about mapping in humans, is now asked routinely of mapping in animals. But the most taken-for-granted tone about animal mapping appears in the most recent issue of the *National Geographic* where Eugene Linden quotes Swiss biologist Christophe Boesch to this effect: "According to Boesch, the chimps maintain a mental map of places where they have left stones from previous sessions. When a panda tree is fruiting, they seem to know the direct route to the nearest stone hammer, often retrieving rocks out of sight more than 200 meters away. Boesch believes that this feat is based on an 'evolved mental map,' a sense of Euclidean space that does not usually occur in a child before the age of nine" (Eugene Linden, "Apes and Humans," *National Geographic Magazine, 181*[3], March, 1992, p. 26). *Euclidean space!*

18. This is not a point ever explicitly made by Humberto Maturana and Francisco Varela, but only because it is subsumed within their definition of behavior in general, namely, "the changes of a living being's position or attitude, which an observer describes as movements or actions in relation to a certain environment." In other words, *living* is a problem of mapping, that of Waterman, for example, being no more than the special case of mobile multicellular sensorimotor correlation. See Maturana's seminal "The Neurophysiology of Cognition" in Paul Garvin, ed., *Cognition: A Multiple View*, Spartan Books, New York, 1970. (In the same volume check out Heinz von Forester's "Thoughts and Notes on Cognition" for more on the organismic significance of mapping.) The most recent synthesis of Maturana's biological phenomenology is his and Francisco Varela's *The Tree of Knowledge*, *op. cit.* In this framework human life without the ability to make mental maps is simply . . . *inconceivable*.

19. That is, mental mapping as we know it today appeared when our brain as we know it today appeared. When was this? It is widely accepted that anatomically modern *Homo sapiens* originated in Africa between 100,000 and 400,000 years ago. See C. B. Stringer and P. Andrews' 1988 "Genetic and Fossil Evidence for the Origin of Modern Humans," *Science*, *239*, 1263–1268, and, for a slightly different perspective, "The Multiregional Evolution of Humans," *Scientific American*, April, 1992, pp. 76–83. In answer to the more specific "When does the modern *brain* evolve?" Philip Lieberman is to the point: "The reorganization that makes voluntary control of speech possible is one of the defining characteristics of the modern human brain. It undoubtedly had occurred 100,000 years ago in anatomically modern fossil hominids as Jebel Qafzeh and Skhul V, who had modern human vocal tracts . . . The archeological evidence associated with Jebel Qafzeh and Skhul V is consistent with their possessing a fully modern human brain—one adapted for complex syntax and logic—but an earlier origin cannot be ruled out. Although much of the enlargement of the prefrontal cortex may derive from the specific contributions of speech and thought, it enters into virtually all aspects of behavior. Therefore any cognitive activity that enhanced biological fitness could have contributed to its development." Given the presence of mapping skills in phylogenetically prior organisms such as birds, fish and bees, can it be doubted that mapping was one of these activities? After all, the brain Lieberman describes as coming into being 100,000 years ago is *our* brain, one manifestly capable of mental mapping. See Lieberman's recent, *Uniquely Human: The Evolution of Speech, Thought, and Selfless Behavior*, Harvard University Press, Cambridge, 1991, pp. 109–110. For a radically different approach to the problem based on an analysis of stone artifacts, see Thomas Wynn's *The Evolution of Spatial Competence*, *Illinois Studies in Anthropology No. 17*, University of Illinois Press, Urbana, 1989. Using an explicitly Piagetian framework, he unequivocally states, "The most straightforward, and perhaps the most unexpected, consequence of this Piagetian analysis is that hominids had achieved operational intelligence by 300,000 years ago, and perhaps earlier" (p. 89).

20. I have argued for this recapitulation, at least with respect to graphic representations in two papers, "Now and Then: Comparisons of Ordinary Americans' Symbol Conventions with Those of Past Cartographers," *Prologue: The Journal of the National Archives*, 9(3), Fall, 1977, pp. 151–161, and "Cultured

Symbols: Thoughts on the Cultural Context of Cartographic Symbols," *Cartographica, 21*(4), Winter 1984, pp. 9–37. The latter article explains what the former describes, but see Chapter Six for my most recent thinking on the topic.

21. Which is not to say that expressions of these maps will be identical, anymore than the universal ability to speak has led to a single language.

22. Robert Beck and I have long stressed the significance of microgenetic processes. See Robert Beck and Denis Wood, "Cognitive Transformation of Information from Urban Geographic Fields to Mental Maps," *Environment and Behavior,* 8(2), June, 1976, and "Comparative Developmental Analysis of Individual and Aggregated Cognitive Maps of London," in Gary Moore and Reginald Golledge, editors, *Environmental Knowing,* Dowden, Hutchinson and Ross, Stroudsburg, Pa., 1976; and more recently, Denis Wood and Robert Beck, "Janine Eber Maps London," *op. cit.*

23. G. Malcom Lewis, "The Origins of Cartography," in J. B. Harley and David Woodward, eds., *The History of Cartography: Volume One: Cartography in Prehistoric, Ancient, and Medieval Europe and the Mediterranean,* University of Chicago Press, Chicago, 1987, p. 50. In "Semantics, Symbols, Geometries and Metrics: Causes of Misunderstanding in the Cartographic Communication of Geographical Information Between Cultures," a paper presented at the Annual Meeting of the American Association of Geographers, April, 1992, San Diego, Lewis disavows his former position, adopting one in most respects identical to that presented here. Catherine Delano Smith's article in the Harley and Woodward volume is far more problematic, for it was clear from the beginning that Lewis was actually talking about *consciousness* rather than cognition, but what Smith is talking about is entirely unclear. See my review article of Harley and Woodward in *Cartographica, 24*(4), Winter, 1987, pp. 69–78.

24. David Woodward, "Reality, Symbolism, Time, and Space in Medieval World Maps," *Annals of the Association of American Geographers, 75*(4), 1985, pp. 510–521.

25. And just because they get made doesn't mean they get published, as Barbara B. Petchenik reminds us: "Intellectual power and intent are important and interesting—but *the maps that actually get made are maps that individuals or groups are willing to pay for*" (personal communication, April 16, 1992).

26. Heinz Werner, *Comparative Psychology of Mental Development,* International Universities Press, New York, 1948. Werner's idea of development is extremely powerful, but his treatment of "primitive man" is problematic. Unlike Lewis and Smith he does not deny them operational capacity: "So far as the primitive man carries out technical activities in space, so far as he measures distances, steers his canoe, hurls his spear at a certain target, and so on, his space as a field of action, as a pragmatic space, does not differ in its structure from our own" (p. 167); but he does qualify this space once it is made a subject of representation and reflective thought (where Werner's use of "representation" is not Piaget's): "The idea of space for primitive man, even when systematized, is syncretically bound up with the subject" (p. 167). I am not absolutely certain what to make of this, but certainly he is speaking here of what others would refer to as "consciousness" or even "world view", both of which are assumed to have deep connections with lifeway in general, that is, with economic structure and social organization (that is, with what some economists might refer to as the

level of development). All theorists concur that this nexus of forces—shorthanded as "hunting and gathering economy" or "feudalism" or "post-Fordian capitalism"—has a decisive effect on the way those who reproduce it "see the world", that is, they concur that less developed societies conceptualize their relationship to the world in a different way than more developed ones do. At this point it doesn't matter whether we think of this difference as one between a Buberian "I–Thou" and "I–It", or an Eliadean "sacred" and "profane," or a Connertonian "incorporating" and "inscribing," or the Wernerian "syncretic" and "discrete" (or alienated to put a less positive face on it), so much as we acknowledge that in this we are not discussing intelligence or cognitive ability but the consciousness engendered by a mode of production, that is, by a . . . *living*. The development of this consciousness can then take its rightful place within the theoretical structures used to describe the transformations of productive systems—as in Henri Lefebvre's *The Production of Space* (Blackwell, Oxford, 1991)—rather than in those used to develop the biology of the brain (that is, refocus our attention on anthropology, history and sociology instead of neurophysiology and psychology).

27. Jean Piaget and Bärbel Inhelder, *The Child's Conception of Space*, Routledge and Kegan Paul, London, 1956. Written in 1948, this is the keystone text for the study of the development of spatial cognition. Also see Monique Laurendeau and Adrien Pinard, *The Development of the Concept of Space in the Child*, International Universities, Press, New York, 1970.

28. Published as "Centering of Mental Maps of the World," *National Geographic Research* 4(1), 1988, pp. 112–127.

29. The handiest general introduction to the project must be Victoria Bricker and Gary Gossen, eds., *Ethnographic Encounters in Southern Mesoamerica: Essays in Honor of Evon Zartman Vogt, Jr., Studies on Culture and Society* 3, Institute for Mesoamerican Studies, The University at Albany, State University of New York, 1989. In my treatment of the Zincanteco here I draw on my own experience in the field as well as on an extensive monographic literature. Especially important are E. Z. Vogt's *Zinacantan: A Mayan Community in the Highlands of Chiapas* (Harvard University Press, Cambridge, 1969); Frank Cancian's *Economics and Prestige in Mayan Community* (Stanford University Press, Stanford, 1965); Gary Gossen's *Chamulas in the World of the Sun: Time and Space in a Maya Oral Tradition* (Harvard University Press, Cambridge, 1974); George Collier's *The Fields of the Tzotzil* (The University of Texas Press, Austin, 1975); and Robert Wasserstrom's *Class and Society in Central Chiapas* (University of California Press, Berkeley, 1983). As the only one to treat the problem of *history*, Wasserstrom's book is particularly valuable, yet despite his continuously critical tone, he reaches conclusions quite close to those of the Harvard Chiapas Project workers, especially Collier.

30. Denis Wood, "Mapping and Mapmaking," paper read at the Annual Meeting of the Association of American Geographers, April, 1992, San Diego.

31. These are hot topics today, but I still turn to Marshall McLuhan's *The Guttenberg Galaxy* (University of Toronto Press, Toronto, 1962).

32. Rundstrom, *op. cit.*, p. 165.

33. J. B. Harley, "Victims of a Map: New England Cartography and the Native Americans," (paper read at The Land of Norumbega Conference, Port-

land, Maine, 1988, p. 17), quoting William Cronon (*Changes in the Land: Indians, Colonists, and the Ecology of New England*, Hill and Wang, New York, 1983, p. 66).

34. David Turnbull, *Maps Are Territories: Science Is an Atlas*, Deakin University Press, Geelong, Victoria (Australia), 1989, p. 42. This is a wise and beautiful book. See my, "Maps Are Territories/Review Article," *Cartographica*, 28(2), Summer, 1991, pp. 73–80.

35. Though it's not as though, *within the Western tradition*, we don't map things that are, from its own perspective, every bit as . . . what? Mystical? For example, *ley-lines*, purportedly ancient pathways or forms of energy used by prehistoric peoples but since lost to modern knowledge. See Claire Cooper Marcus, "Alternative Landscapes: Ley-Lines, Fêng-Shui and the Gaia Hypothesis," *Landscape*, 29(3), 1987, pp. 1–10. The great Western tradition is permitted to retain its appearance of impeccable rationality only by ignoring, suppressing and/or denying its mystical, its religious, its irrational . . . its *Dionysian* side.

36. Again, what has been integrated here have been not only the topo sheets, property maps, congressional district maps and so on, but the maps of the ley-liners, the Flat Earth Society, those who believe in a hollow earth, and the like; to say nothing of the volumes of maps illustrating novels and games of fantasy and science fiction, *ad infinitum*.

37. Stephen Hall, *Mapping the Next Millennium*, Random House, New York, 1992.

38. Again, mapmaking may *originate* for navigational purposes—as among the Micronesians—or to represent the "footprints of the Ancestors"—as among Aboriginal-Australians, or as cadasters—as (perhaps) among the Egyptians and Babylonians—but it does not seem to *develop*, that is, undergo differentiation, articulation and hierarchic integration, in the absence of a need to keep records. I am arguing that such development only occurs when societies *grow* to the point where they need to keep records; and then record keeping pushes the development of systems of representation; out of which mapmaking as a *lifeway* develops as such systems of representation become differentiated. This means that more societies will engage in mapping than will become mapmaking societies.

39. See Denise Schmandt-Besserat, "An Archaic Recording System and the Origin of Writing," *Syro-Mesopotamian Studies*, *1*(2), July, 1977, pp. 1–70. Her most recent synthesis appears as *Before Writing: From Counting to Cuneiform, Vol. 1*, University of Texas Press, Austin, 1991. In his review of this book John Alden concludes by observing that, "The evidence presented in this book supports an important general principle that is surely familiar to readers of this magazine. Human inventions, like biological adaptations, rarely appear out of the blue. Instead they typically involve the modifications of an existing device or structure to serve an expanded function" (*Natural History*, March 1992, p. 67). My point exactly.

40. Mary Elizabeth Smith, *Picture Writing From Ancient Southern Mexico: Mixtec Place Signs and Maps*, University of Oklahoma Press, Norman, 1973.

41. There are many examples in Smith, *ibid.*, and one illustrates the Property Map entry in Helen Wallis and Arthur Robinson, *op. cit.*, p. 101.

42. Woodward, *op. cit.* Also see his more exhaustive treatment in "Medieval *Mappaemundi*" in Harley and Woodward, *op. cit.*, 286–370.

43. Jerome Bruner *et al.*, *Studies in Cognitive Growth*, John Wiley & Sons, New York, 1966, p. 2 and p. 6.

44. Brian Harley, "Silences and Secrecy: The Hidden Agenda of Cartography in Early Modern Europe," paper read at the XIIth International Conference on the History of Cartography, Paris, September, 1987, p. 1. He put this slightly differently, but to the same end, in the published version which appeared under the same title in *Imago Mundi*, *40*, 1988, pp. 57–76. But this was increasingly the point Harley was making in the final years of his life, that cartography *was* political discourse in the service of the state.

45. In addition, of course, to the acknowledged stick-chart, *dhulan*, bark painting, and so on. The point, to reiterate, is not that these peoples do not make maps, but that the maps they make remain relatively isolated, unusual, special, rare—in contrast to the taken-for-granted character of maps in a mapmaking, that is, *map-immersed*, society.

46. Katharine Milton, "Civilizations and Its Discontents", *Natural History*, March 1992, p. 38.

47. *Ibid.*, p. 39.

48. *Ibid.*

49. Actually the most eloquent testimony comes from the André Kertész collection of photographs entitled *On Reading* (Grossman, New York, 1971), but it was Marshall McLuhan who first characterized it as a problem of consciousness (in *The Guttenberg Galaxy*, *op. cit.*). For McLuhan it was type—and reading—that silenced the human *voice*, producing the alienated consciousness we have today, making the modern state possible no less than the map: "print created national uniformity and government centralism, but also individualism and opposition to government as such" (p. 235). To reiterate the points made with respect to Werner in Note 26, what is at stake with respect to the differences that exist between our mapmaking (and print reading) society and the oral societies Milton describes is not differences in fundamental cognitive abilities, but in the *consciousnesses* that different lifeways inevitably produce. For a more focused attack on the impact on consciousness resulting from the reading (and writing) of printed books, read Alvin Kernan's completely convincing *Printed Technology, Letters and Samuel Johnson* (Princeton University Press, Princeton, 1987).

50. Joseph Conrad, *Heart of Darkness*, The Heritage Press, Norwalk, Connecticut, 1969, p. 10. It's worth noting that the novel was written in 1898–1899 out of Conrad's experiences of 1890.

51. J. K. Wright, "*Terrae Incognitae:* The Place of the Imagination in Geography," in his *Human Nature in Geography: Fourteen Papers, 1925–1965*, Harvard University Press, Cambridge, 1966, p. 68.

52. *Ibid.*, p. 69.

53. J.B. Harley, "Victims of a Map" *op. cit.* p. 22

54. Harley, *ibid.*, reviews the literature, but see also the work of G. Malcom Lewis who has made this scholarly territory his own: "The Indigenous Maps and Mapping of North American Indians," *The Map Collector*, 9, 1979, pp. 25–32; "Indicators of Unacknowledged Assimilations from Amerindian *Maps* on Euro-American Maps of North America," *Imago Mundi*, *38*, 1986, pp. 9–34; and "Native North Americans' Cosmological Ideas and Geographical Awareness,"

in John Allen, editor, *North American Exploration*, University of Nebraska Press, Lincoln, forthcoming.

55. This also is Conrad speaking through Marlowe: "The conquest of the earth, which mostly means taking it away from those who have a different complexion or slightly flatter noses than ourselves, is not a pretty thing when you look into it too much." (Conrad, *op. cit.*, p. 8). My point is simply that this taking away is all but complete when the taken get caught up on the takers' maps.

56. Harley, "Victims of a Map", *op. cit.*, p. 1.

57. It's the eye of Sauron: "In the black abyss there appeared a single Eye that slowly grew, until it filled nearly all the Mirror. So terrible was it that Frodo stood rooted, unable to cry out or to withdraw his gaze. The Eye was rimmed with fire, but was itself glazed, yellow as a cat's, watchful and intent, and the black slit of is pupil opened on a pit, a window into nothing. Then the Eye began to rove, searching this way and that; and Frodo knew with certainty and horror that among the many things that it sought he himself was one" (J.R.R. Tolkien, *The Fellowship of the Ring*, Houghton, Mifflin, Cambridge, 1954, p. 379). Tolkien is an unsung phenomenologist of the panopticon.

58. There is not just one story like this. The Chinese too were a mapmaking people, and so were the Arabs. On a less global scale the Aztecs and Incas put other Amerindian groups on their maps. Perhaps it would be better to think of every people as a mapmaking people, varying in the extent to which they included other peoples on their maps. But it would be disingenuous to pretend that the difference between our mapping of the Mayoruna realm and their not mapping that of the United States amounted to no more than a *quantitative* distinction. They are in our orbit, as we are not in theirs.

59. Nor did this happen only in the past. The expropriation of land and cultural capital from indigenous Americans continues unabated in the rainforests of Brazil and Venezuela, as well as elsewhere.

60. Orlando and Claudio Villas Boas, *Xingu: The Indians, Their Myths*, Farrar, Straus, Giroux, New York, 1973.

61. For a more detailed treatment of this way of distinguishing "the West" from such smaller societies, see my review in *Environment and Behavior* (September 1985, pp. 643–647) of Rix Pinxten, Ingrid van Dooren and Frank Harvey's *The Anthropology of Space*, University of Pennsylvania Press, Philadelphia, 1983.

Chapter Three

1. Gilbert Grosvenor, "New Atlas Explores a Changing World," *National Geographic*, 178(5), November, 1990, pp. 126–129.

2. *National Geographic Atlas of the World, Sixth Edition*, National Geographic Society, Washington, D.C., 1990, p. v.

3. A *Clear Day*, poster published by Northern Telecom, 1992, to celebrate their becoming "the first global telecommunications company to eliminate chlorofluorocarbon (CFC-113) solvents from its manufacturing operations."

4. Pamela E, Mack, *Viewing the Earth: The Social Construction of the Landsat Satellite System*, MIT Press, 1990, p. 39.

5. Roland Barthes, "The Photographic Message," in *Image-Music-Text*, Hill and Wang, New York, 1977, pp. 16–17.

6. *National Geographic Atlas*, *op. cit.*, p. v.

7. The 35 million figure comes from the *National Geographic Atlas*, *ibid.*, but according to Van Sant's own "GeoSphere Project Report," October, 1991, the figure is actually 37.3 million. Does it matter? The number is enormous in either case . . .

8. Stephen Hall, *Mapping the Next Millennium: The Discovery of New Geographies*, Random House, New York, 1992, p. 62.

9. Or something. "Technically," writes Mack, "Landsat produces images not photographs (a photograph is defined as an image formed directly on a recording medium by electromagnetic radiation from the subject, thus excluding even television cameras). *Picture* is used here as a nontechnical synonym for *image*. Many technical people object to calling Landsat images pictures because picture may be seen as a synonym for photograph (there is a particular concern to differentiate Landsat data from the photographs taken by reconnaissance satellites). It is important to remember that Landsat data could be prepared for study not only in the form of images reproduced on photographic film but also in the form of computer tapes" (Mack, *op. cit.*, p. 215). As we will see, Van Sant's map was constructed from TIROS, not Landsat, satellite imagery, but the similar imaging systems were pioneered on Landsat and identical strictures apply.

10. The early history of weather satellites is best told by Richard L. Chapman in his doctoral dissertation, *A Case Study of the U. S. Weather Satellite Program: The Interaction of Science and Politics* (Syracuse University, Syracuse, 1967). But also see the relevant NOAA technical memoranda: Arthur Schwalb, *Modified Version of the Improved TIROS Operational Satellite (ITOS D-G)*, U. S. Department of Commerce, NOAA Technical Memorandum NESS 35, Washington, D. C., 1972; Arthur Schwalb, *The TIROS-N/NOAA A-G Satellite Series*, U. S. Department of Commerce, NOAA Technical Memorandum NESS 95, Washington, D. C., 1978; Levin Lauritson *et al.*, *Data Extraction and Calibration of TIROS-N/NOAA Radiometers*, U. S. Department of Commerce, NOAA Technical Memorandum NESS 107, Washington, D. C., 1979; and for an overview Dennis Dismachek *et al.*, *National Environmental Satellite Service Catalogue of Products, Third Edition*, U. S. Department of Commerce, NOAA Technical Memorandum NESS 109, Washington, D. C., 1980.

11. Mack, *op. cit.*, p. 35.

12. *Ibid.*, p. 74.

13. Mack, *op. cit.*, pp. 74–75, quotes a retired air force general as advising NASA that, "in the Intelligence Community, at the working levels at least, there is the traditional, understandable feeling that almost any earth sensing from satellites by NASA, with their inevitable public disclosure, will be harmful, and perhaps fatal, to classified sensing for national security purposes." Only very recently, with the signing reported *in this morning's paper* by George Bush of a directive clearing the way for environmental scientists to use spy satellites (planes, ships and submarines) and records, has this attitude begun to change

(William Broad, "Global-warming sleuths get peek at Cold War spy data," *News and Observer*, June 23, 1992, p. 7B).

14. Mack, *op. cit.*, p. 75.

15. *Ibid.*, pp. 76–77.

16. *Ibid.*, p. 79. Of course what is all this wrangling but the very way in which our society establishes the priorities on which it intends to act? In each of these compromises over the scanner, over the spectral windows—we sense the emergence of the author, or at least of an authorial voice, which speaks through the map, not just Van Sant's, but that of the whole society which has conspired to make *possible* the map he will finally make.

17. Van Sant's images came from the Advanced Very High Resolution Radiometer which scans in one (or two) visible and three (or four) infrared channels. The number depends on the individual satellite: TIROS-N/NOAA-A, -B, -C and -E satellites carry four-channel instruments; TIROS/NOAA-D, -F, and -G satellites, are five-channel instruments.

18. *Ibid.*, p. 108. Mack's argument here is quite subtle and I do not do it justice, but explicitly political considerations color every decision made throughout the Landsat and other satellite mapping programs.

19. The French SPOT satellite dispenses with the mirror and prism of the Norwood [that is, the Hughes Aircraft (that is, General Motors)] multispectral scanner. Instead it deploys multilinear array sensors, but to the same end.

20. Mack, *op. cit.*, p. 47. For remote sensing in general, go to the horse's mouth: Robert Colwell, editor, *Manual of Remote Sensing, Second Edition*, American Society of Photogrammetry, Falls Church, Va., 1983.

21. *National Geographic Atlas of North America: Space Age Portrait of a Continent*, National Geographic Society, Washington, D.C., 1985, p. 7. The subtitle is not irrelevant, touching as it does on the theme of the map as portrait, especially when it involves, as this atlas famously does, satellite imagery.

22. Technically we should probably think about this as a *re*coding since what is *green* = *vegetation*, *blue* = *water* and so on but a code, one whose decipherment we think about as . . . *learning to see?*

23. *National Geographic Atlas of the World*, *op. cit.*, p. v.

24. Harold Osborne, editor, *The Oxford Companion to Art*, Oxford University Press, Oxford, 1970, p. 955.

25. Paul Edwards, editor, *The Encyclopedia of Philosophy*, Macmillan, New York, 1967, Vol. 7, p. 77.

26. Alex Preminger, editor, *Princeton Encyclopedia of Poetry and Poetics*, Princeton University Press, Princeton, 1974, p. 685.

27. Waldo Tobler, "Transformations," in William Bunge, editor, *The Philosophy of Maps*, Michigan Inter-University Community of Mathematical Geographers, Discussion Paper No. 12, Ann Arbor, June, 1968.

28. William Bunge, *Theoretical Geography*, Lund Studies in Geography, Series C: General and Mathematical Geography, No. 1, University of Lund, Lund, Sweden, 1962, p. 216.

29. Waldo Tobler, "A Classification of Map Projections," *Annals of the Association of American Geographers*, 62, 1962, p. 167.

30. Arthur Robinson *et al.*, *Elements of Cartography: Fifth Edition*, John Wiley & Sons, New York, 1984, p. 79.

31. Edward Espenshade and Joel Morrison, editors, *Rand McNally Goode's World Atlas, 16th Edition*, Rand McNally, Chicago, 1982, p. x.

32. Wellman Chamberlin notes, "When you read one of the many current articles on cartography which begin by abolishing Mercator, stop to realize that the Mercator projection is standard for sea navigation all over the world and is constantly used for aeronautical charts" and so forth and so on. He makes the remark in his *The Round Earth on Flat Paper: Map Projection Used by Cartographers*, National Geographic Society, Washington, D.C., 1947, p. 82.

33. *Ibid.*, p. 72. A similarly mutually exclusive choice must be made between showing consistent scale along one or more standard lines *or* in all directions but from only one or two points. It is also impossible to represent all directions correctly. The authority here remains Derek Maling, *Coordinate Systems and Map Projections*, George Philip, London, 1973.

34. John Garver, "New Perspectives on the World," *National Geographic*, *174*(6), December, 1988, p. 913.

35. Arthur Robinson, "Arno Peters and His New Cartography," *American Cartographer, 12*(2), 1985, pp. 103–111.

36. Mark Monmonier, *How to Lie with Maps*, University of Chicago Press, Chicago, 1991, pp. 172–174.

37. I don't think he did, but make up your own mind. See Arno Peters, *Peters Atlas of the World*, Harper and Row, New York, 1990.

38. Robinson, *op. cit.*, claimed the continents on the Peters projection looked like, "wet, ragged, long winter underwear hung out to dry on the Arctic Circle." *This* is the voice of science!

39. Monmonier, *op. cit.*, p. 97. Also see Peter Vujakovic, "Arno Peters' cult of the 'New Cartography': from Concept to World Atlas," *Bulletin of the Society of University Cartographers*, *22*(2), 1989, pp. 1–6.

40. Hall, *op. cit.*, p. 380.

41. *Ibid.*,p. 380–381.

42. David Turnbull, *Maps Are Territories: Science Is an Atlas*, Deakin University Press, Geelong, Victoria (Australia), 1989, p. 7.

43. Hall, *op. cit.*, p. 382.

44. This is another great alibi cartographers whip out whenever necessary. See my review-essay of David Woodward, editor, *Art and Cartography: Six Historical Essays* (University of Chicago Press, Chicago, 1987), and his reply, both in *Cartographica*, *24*(3), Autumn, 1987, pp. 76–85. On the social construction and historical contingency of aesthetics in general see Pierre Bourdieu, *Distinction: A Social Critique of the Judgment of Taste*, Harvard University Press, Cambridge, 1984; and Terry Eagleton, *The Ideology of the Aesthetic*, Blackwell, Oxford, 1990.

45. Garver, *op. cit.*, p. 913.

46. Actually, it looks . . . *shocking*, though only to those who have been confusing the map and the globe. The trade-off is that it has other attributes than its preservation of areal equivalency that make it peculiarly easy to use. For example, it maintains east–west and north–south directions too.

47. J. B. Harley, "Can There Be a Cartographic Ethics?" *Cartographic Perspectives*, *10*, Summer, 1991, pp. 10–11. ACSM stands for American Congress on Surveying and Mapping.

48. This, by the way, is not a straightforward operation. As Robert Richardson pointed out in a technical note, the projection "is unusual in having no definition by mathematical formula" ("Area Deformation on the Robinson Projection," *The American Cartographer, 16*[4], October, 1989, p. 294), a deficiency partially rectified by John Synder's publication of a computation algorithm only recently ("The Robinson Projection—A Computation Algorithm," *Cartography and Geographic Information Systems, 17*[4], October, 1990, pp. 301–305). Only the *National Geographic* used the Robinson to project the Van Sant. Northern Telecom used another, unidentified projection for its poster, one that replaced the curved outline of the Robinson with one more rectangular.

49. Grosvenor, *op. cit.*, pp. 126–27.

50. W. T. Sullivan, *Earth at Night*, Hansen Planetarium, Salt Lake City, 1986. When ordered from the planetarium, the poster is accompanied by a sheet of further information replete with details about the sensors, a bibliography, and "points to Ponder and Questions to Pursue." No bones are made about the fact that this is a political tract concerned with light pollution of the night sky.

51. Indeed it raises the whole question about resolution again. Recall that L. Gordon Cooper could see the human landscape, terrifyingly clearly. The loss of resolution produced by military paranoia has its payoff here in a landscape so crudely rendered that the human touch can all but not be seen under daylight conditions (this will relieve many).

52. Again, the map *always* shows us something that can't be seen, a reality that exceeds our vision. That's precisely its point, precisely what makes it valuable. But that necessarily renders it the product of a social construction, and this prevents it from assuming the mantle of the Natural, that is, of the naturally true.

53. Roger H. Ressmeyer, "The Moon's Racing Shadow," *National Geographic, 181*(5), May, 1992, pp. 38–39. The point of the four images was to catch the path of the moon's shadow, a dark band girdling the earth. It's a powerful and provocative image.

54. *National Geographic Atlas of the World, op. cit.*, p. v.

55. James Underwood Crockett, *Bulbs*, Time-Life Books, New York, 1971, pp. 148–149.

56. But their indexicality is all *toward* this duration which they then, precisely, fill with the movement of things in space. The Van Sant points toward no duration, suppresses, in its image, any suggestion of dynamism, change, effort, movement. In fact, these are all displaced . . . *to the map authors*. The earth, like a virgin bride, remains . . . *untouched*. For some cogent remarks on this issue see David Woodward, "Reality, Symbolism, Time, and Space in Medieval World Maps," *Annals of the Association of American Cartographers, 75*(4), 1985, especially pp. 519–520.

57. Grosvenor, *op. cit.*, p. 127.

58. Such decisions are among the most fundamental an author—or an artist—makes. It can never be forgotten that whatever the appearances of automatically emerging from the print dryer, this map was *authored*, though, again, by analogy with filmmaking, *directed* might be a better choice of word.

59. Barthes understood perfectly well that the photograph was coded, and not only at the sensory level. "This purely 'denotative' status of the photograph,

the perfection and plenitude of its analogy, in short its 'objectivity,' has every chance of being mythical (these are the characteristics that common sense attributes to the photograph). In actual fact, there is a strong possibility (and this will be a working hypothesis) that the photographic message too—at least in the press—is connoted" (Barthes, *op. cit.*, p. 19).

60. *Ibid.*, p. 16.

61. *Ibid.*, p. 20.

62. *Ibid.*, p. 25.

63. *Ibid.*, p. 21

64. *Ibid.*, p. 23.

65. Grosvenor, *op. cit.*, p. 126.

66. *National Geographic, 174*(6), December, 1988, cover and pp. v, xxvii (neither paginated), and 765.

67. Daniel J. Kevles, "Some Like It Hot," *The New York Review*, March, 26, 1992, p. 33.

68. Woodward, *op. cit.*, p. 519.

Chapter Four

1. John Van Pelt, "The Loose World of Mapmaking," *Christian Science Monitor*, September, 18, 1991.

2. *Ibid.*

3. The 1:100,000,000 is the approximate scale of most of the world thematic maps in Edward Espenshade and Joel Morrison, editors, *Rand McNally Goode's World Atlas, 16th Edition*, Rand McNally, Chicago, 1982.

4. As Van Sant was interested in the land and water surface but not in the clouds, as he was interested in the earth by day and not by night. Of course we want to refer *through* these local interests to more global interests too, the need to address the condition of the planet as a whole in Van Sant's case.

5. This scarcely exhausts the messages of this map which, published as a promotional freebie by McCormick in 1955, says as much about America in the Fifties as any other single document I've seen. While "speaking" of "The Romance of Spices" and the history of cartography (shamelessly appropriated), in its brilliant and subtle way the map also says "America, heritor and guardian of Western civilization, brings the world to its citizens who sip it in their tea." In so many words: "The long journey down great mountains and rivers, through distant cities and over strange oceans to your teapot or cup is ended. Let's pour another cup of fragrant, cheering, refreshing McCORMICK TEA!" The map merits the monographic analysis ordinarily devoted to 16th and 17th century decorative printed maps in whose tradition it very self-consciously situates itself.

6. Published by Archar (Toronto, Canada) in 1981, the map refers to itself as a "City Character Print," but I agree with the Southworths (Michael and Susan Southworth, *Maps*, Little, Brown, Boston, 1982) who call it a "Floating Landmark Map," noting that "The 'floating landmark' style is popular for tourist and poster maps, which are essentially montages of key images. Precise information is unnecessary and distortion is a crucial part of the style. Landmarks are drawn at an enlarged scale and face in any direction without regard for

reality. As a result buildings that are actually quite remote from one another appear to be next-door neighbors. Because of the distortions, omissions, incorrect orientation, and misleading juxtapositions, it is not possible to use this type of map for navigation. Rather, it serves as a poster, souvenir, or advertisement" (p. 88). I especially appreciate their treatment of "distortion" as an aspect of "style," that is, as a *rhetorical device*.

7. John Garver, "New Perspectives on the World," *National Geographic*, *174*(6), December, 1988, p. 913.

8. That is, the inherent impossibility of maintaining on a flat piece of paper all the spatial relationships existent on the globe. This little bit of special pleading is always the cartographer's *first* resort, for if he *must* distort the world, then . . . that doesn't *really* count (because unwilled, because required by . . . *the way things are*).

9. Mark Monmonier, *How to Lie with Maps*, University of Chicago Press, Chicago, 1991, p. 1.

10. Roland Barthes, *Mythologies*, Hill and Wang, New York, 1972, p. 11.

11. R. D. Laing, *The Politics of the Family*, Pantheon, New York, 1971, p. 98.

12. *Inter alia:* Hans Speir, "Magic Geography," *Social Research*, 8, 1941, pp. 310–330; Louis Quam, "The Use of Maps in Propaganda," *Journal of Geography*, 42, 1943, pp. 21–32; Louis Thomas, "Maps as Instruments of Propaganda," *Surveying and Mapping*, 9, 1949, pp. 75–81; F. J. Ormeling, "Soviet Cartographic Falsifications," *Military Engineer*, 62, 1970, pp. 389–391; John Ager, "Maps and Propaganda," *Bulletin of the Society of University Cartographers*, *11*, 1977, pp. 1–14; Phil Porter and Phil Voxland, "Distortion in Maps: the Peters Projection and Other Devilments," *Focus*, 36, 1986, pp. 22–30; *ad nauseum*.

13. J. B. Harley, "Maps, Knowledge, and Power," in Dennis Cosgrove and Stephen Daniels, editors, *The Iconography of Landscape: Essays on the Symbolic Representation, Design and Use of Past Environments*, Cambridge University Press, Cambridge, 1988, p. 278.

14. This is in accordance with Barthes' dictum: "The natural is never an attribute of physical Nature; it is the alibi paraded by a social majority: the natural is a legality" (Roland Barthes, *Roland Barthes*, Hill and Wang, New York, 1977, p. 130).

15. Robert Rundstrom, "A Cultural Interpretation of Inuit Map Accuracy," *Geographical Review*, 80(2), p. 162.

16. *Ibid.*, p. 156.

17. In other words, those elements of society that use the map to subdue the land to their purposes deny both their own and the map's role in this process ("nobody subdued anything and the map just recorded things the way they were") in order to reap the rhetorical (and political) advantages of passing the status quo off as natural, that is to say, as inevitable (so why resist?).

18. Morris M. Thompson, *Maps for Americans: Cartographic Products of the U.S.Geological Survey and Others*, Department of the Interior, Washington, D.C., 1979, p. 27 (emphasis added).

19. But since it's permanent, we recognize the oxymoron, *permanent change*.

20. *Ibid.*, p. 28. Another coding is implied here too: permanent = cheap (because you don't have to revise the maps so much). As we'll see, whenever you focus on one of these issues . . . *it turns into the other*. Bill Bunge has added this:

"Considering the enormous change on the earth's surface, it must be constantly re-surveyed. Man is the only species who not only has revolutionized the biological landscape but has made actual dents into the geomorphology and climatology of the planet. The most important change in the earth's landscape is not any shift that would be perceivable on an aerial photograph, it is the *shift in what we value*" (W. Bunge, *Detroit Geographical Expedition, Field Notes, 1*, 1969, p. 2). Amid such change, what is there to be . . . *permanent?*

21. *Ibid.*, p. 29. Though the legend is not printed on the maps—it is not small—it is readily available in pamphlets and brochures.Variations appear on older and younger maps. Thompson also includes a chart of "features shown on topographic maps" (pp. 115–121) which indicates the potential presence of 213 items including "area to be submerged," "ditch, intermittent," "lake or pond, intermittent" and other "permanent" features.

22. Martin Taylor, *et al.*, "Mack Avenue and Bloomfield Hills—From a Child's Point of View," in *Geography of the Children of Detroit, Detroit Geographical Expedition and Institute Field Notes, 3*, 1971, between pp. 19 and 20.

23. You could map it just about any way you wanted to. Each toy or piece of trash need be shown no more than every tree beneath the woodland tint is shown or every building beneath the "bldg. omission area" tint. A yellow tint could stand for "Areas friendly to children," gray for "Areas hostile to children." This would result in the sort of map Bunge imagined in his *Atlas of Love and Hate*. See the map, "Grassless Space—The Karst," which shows in four categories the percentage of grassless lots for a good chunk of downtown Detroit, in W. W. Bunge and R. Bordessa, *The Canadian Alternative: Survival, Expeditions and Urban Change*, York University, Atkinson College, *Geographical Monographs, 2*, 1975, p. 290.

24. Thompson, *op. cit.*, p. 27

25. Monmonier, *op. cit.*, p. 1.

26. This reality is illustrated again and again in Pamela Mack's history of the Landsat satellite system (*Viewing the Earth: The Social Construction of the Landsat Satellite System*, MIT Press, Cambridge, 1990).

27. Thompson, *op. cit.*, p. 27.

28. Sherre Glover, "Speedway tolerable," *News and Observer*, May, 18, 1992, p. 8A. This was a letter to the editor. Noise—from airports, highways, concert venues—is a problem taking up increasing space in this paper. In the local section there was a story about a neighborhood suit against the airport over noise and on the comics page the Canadian strip *For Better or for Worse* continued its story about Phil and his battle with the Bullard's dog next door. "It's not fair, Geo," he says against a frame all but filled with "BARK, BARK, BARKs," "You buy a house for peace and privacy!/A man's home is his castle!/And we have a dragon next door."

29. Bunge and Bordessa, *op. cit.*, p. 16. I demonstrated in 1969 that Mexican teenagers could differentiate time of day and weekday from weekend on the basis of sounds (Denis Wood, "The Image of San Cristobal," *Monadnock, 43* (1969), pp. 24–45); by 1971 I extended this work to show that they could also tell one neighborhood from another on this basis alone (Denis Wood, *Fleeting Glimpses: Adolescent and Other Images of That Entity Called San Cristobal las*

Casas, Chiapas, Mexico, Clark University Cartographic Laboratory, Worcester, Massachusetts, 1971, pp. 167–174).

30. Kevin Lynch, *Site Planning*, MIT Press, Cambridge, 1962. By the second edition, noise and sound merited six entries in the index (*Site Planning, Second Edition*, MIT Press, Cambridge, 1971). The third edition resulted in nine entries, including a seven page appendix on noise (Kevin Lynch and Gary Hack, *Site Planning, Third Edition*, MIT Press, Cambridge, 1984).

31. Kevin Lynch, *Managing the Sense of a Region*, MIT Press, Cambridge, 1976, p. 8.

32. A geographer concerned with this richer world is Douglas Porteous. Chapter 3 of his *Landscapes of the Mind: Worlds of Sense and Metaphor* (University of Toronto Press, Toronto, 1990) is devoted to the soundscape (Chapter 2 to the smellscape).

33. R. Murray Schafer, *The Tuning of the World: Toward a Theory of Soundscape Design*, University of Pennsylvania Press, Philadelphia, 1980.

34. Bunge and Bordessa, *op. cit.*, pp. 17A–B. They skewer the notion that these groves could be inferred from airphotos when they observe that, "Today churches are places of organized activity designed to compete with the hustling environment instead of quiet retreats" (and of course the quiet groves of the schoolyards are time dependent: mapping them would require a notation like that the Survey uses for "land subject to controlled inundation" to mark the periodic inundation . . . *of sound*). The term "quiet groves" is due to Schafer, *op. cit.*, p. 253.

35. Lynch, *op. cit.*, p. 145. The original source of this map was Michael Southworth's *The Sonic Environment of Cities* (unpublished MIT thesis, 1967). This work was summarized and the map reproduced in Southworth's valuable "The Sonic Environment of Cities" (*Environment and Behavior, 1*(1), June 1969, pp. 49–70). Lynch then reproduced the map in 1976, Schafer in 1977 (*op. cit.*, p. 265) and Michael and Susan Southworth in 1982 (*op. cit.*, p. 190), where it has been handsomely redrawn. The quote about escaping visual bondage is from Southworth's *Environment and Behavior* article, p. 69.

36. Lynch, *op. cit.*, p. 101. Kerry Dawson is another eager to move "away from an aesthetic based solely on visual attributes, and toward qualities which offer variety in sensual experience." See his, "Flight, Fancy, and the Garden's Song," *Landscape Journal, 7*(2), Fall 1988, pp. 170–75.

37. Schafer, *op. cit.*, pp. 214–215 and 264 and 267. In this morning's paper a headline in the local section reads, "Residents ask compensation for jet noise: suit contends property values hurt." At issue here is precisely the change to which Schafer refers: " 'From their yard, they have had airplanes measured at 93 to 97 decibels,' said James Fuller, a Raleigh lawyer representing the plaintiffs. 'That is barely under what it would sound like if you were at a rock concert standing under the speakers. Basically the outdoors was taken away from them.' At issue is whether the doubling of flights after the opening of the American Airlines hub in June 1987 has hurt the surrounding residents' property values" (Wade Rawlins, "Residents ask compensation for jet noise," *News and Observer*, May 19, 1992, p. B1).

38. Thompson, *op. cit.*, p. 27.

39. Robertson Barrett, "Residents want more walls to contain noise from Beltline," *The News and Observer*, April 20, 1990, p. 1B. They could have objected as well to the refusal to acknowledge hydrocarbon pollution, the impact on wildlife, historic values, safety and so on. The Survey serves the interests of those who control it, that is, who control the country (cars, concrete).

40. "The din gets louder and louder. More than half the low-altitude air space in the United States is allocated to military flights, and the traffic in both civilian and military air space grows. In Goose Bay, Labrador, the culture of the Innu Indians, one of the last hunting and fishing peoples in North America, is about to vanish. The air space over nearly 40,000 square miles of forest inhabited by the Innu is used by German, Dutch, and British military planes flying low-level training missions. The jets fly down the river valleys, where the Indians fish and hunt for caribou and moose. They scream overhead at treetop level, and the noise leaves the Indians deafened, dizzy, and demoralized. There are currently 8,000 such flights a year, but by the year 1992 there are expected to be 40,000. Says John Olthuis, attorney for the Innu, 'They will be wiped out as a people.' " (Peter Steinhart, "Eavesdropping in the Wilds," *Audubon, 91*[6], November 1989, p. 26.)

41. Thompson, *op. cit.*, p. 27. In a way this all sounds so self-evident it goes without saying, but it is precisely *there*—in the *goes without saying*—that every sort of problem roots. In *The Visual Display of Quantitative Information* (Graphics Press, Cheshire, Connecticut) Edward Tufte has argued that, "Data-rich designs give a context and credibility to statistical evidence. Low information designs are suspect: what is left out, what is hidden, why are we shown so little? High density graphics help us to compare parts of the data by displaying much information within the view of the eye: we look at one page at a time and the more on the page, the more effective and comparative our eye can be. The principle, then, is: maximize data density and the size of the data matrix, within reason" (p. 168). Not that the topo sheet is data poor—in fact, Tufte points with admiration to the 100 million bits of information the Survey layers onto the average quadrangle—but the data it displays has been selected without undue concern for the resolving power of the eye!

42. And indeed ours does. For example, the boundaries on our quadrangle were compiled from a number of different sources, none of them photographic.

43. Eduard Imhof, *Cartographic Relief Representation*, de Gruyter, Berlin, 1982. His point is that this is not a task that can be automated but one that requires the judgment of geographically and graphically trained cartographers. See my review in the *Association of Canadian Map Libraries Bulletin*, *54*, March 1985, pp. 37–40.

44. It is their awareness of this instrumentality that distinguishes Imhof, Tufte and Jacques Bertin from so many other cartographers: "At their best, graphics are instruments for reasoning about quantitative information," Tufte says (*op. cit.*, p. 9). Bertin is even more specific: "A graphic is not 'drawn' once and for all; it is 'constructed' and 'reconstructed' until it reveals all the relationships constituted by the interplay of the data." This implies that, "A graphic is never an end in itself; it is a moment in the process of decision-making." Bertin insists that, "Cartography is above all a means of data

processing . . . It can serve either in the discovery of characteristics corresponding geographically to a given characteristic or in the discovery of a geographical distribution defined by a given set of characteristics" (*Graphics and Graphic Information Processing*, de Gruyter, Berlin, 1981, pp. 16 and 161; but see also his magisterial *Semiology of Graphics*, University of Wisconsin Press, Madison, 1983). Note the emphasis in these quotes on *instruments*, *decision-making*, *data-processing*, *discovery*. Whatever it is, the map is no longer a . . . picture.

45. I am thinking at the moment of the work George Jenks and his students carried out in the '70s at the University of Kansas (see Theodore Steinke's, "Eye Movement Studies in Cartography and Related Fields," in Patricia Gilmartin, *Studies in Cartography: A Festschrift in Honor of George F. Jenks*, *Cartographica*, 24(2), Summer, 1987, pp. 40–73) but more generally of the whole school of thinking about cartography headed by Robinson and Jenks which conceives of the map *above all else* as a picture.

46. Thompson, *op. cit.*, p. 16.

47. Allan Nevins, *Abram S. Hewitt, With Some Account of Peter Cooper*, Harper and Brothers, New York, 1935, p. 120. I have drawn heavily on Nevins for the history of the New Jersey iron industry, to say nothing of the life of Hewitt.

48. See Cyril Stanley Smith and R. J. Forbes, "Metallurgy and Assaying," in Charles Singer *et al.*, editors, *From the Renaissance to the Industrial Revolution*, *Volume III*, *A History of Technology*, Oxford University Press, Oxford, 1957, pp. 27–71; and H. R. Schubert, "Extraction and Production of Metals: Iron and Steel," in Singer *et al.*, editors, *The Industrial Revolution, Volume IV, A History of Technology*, Oxford University Press, Oxford, 1958, pp. 99–117.

49. Nevins, *op. cit.*, p. 98.

50. *Ibid.*

51. Actually they had mortgaged the property to one Peter Townsend. Peter Cooper—New York inventor, industrialist, and idealist (and founder of the Cooper Union *et cetera et cetera*)—first took over the mortgage but subsequently bought the property at its full appraised value. In turn he sold it to the Trenton Iron Company, that is, to his son, Edward Cooper, and future son-in-law, Abram Hewitt. See Nevins, *op. cit.*, p. 120.

52. *Ibid.*, p. 97.

53. *Ibid.*, pp. 98–99.

54. *Ibid.*, p. 120.

55. *Ibid.* p. 124–125.

56. This interdigitation of the industrial and the bucolic was a pervasive presence on the American scene in the period following the Civil War. See the paintings illustrated in Marianne Doezema's *American Realism and the Industrial Age* (Cleveland Museum of Art, Cleveland, 1980), especially Thomas Anshutz's 1896 *Steamboat on the Ohio* (cover, and pp. 28, 29 and 33).

57. Henry Adams, *The Education of Henry Adams*, *An Autobiography*, The Heritage Press, Norwalk, 1942, p. 275. He all but gave up the iron and steel business once he had made his fortune and turned to public service. He served five terms in the U. S. House and was an influential reform mayor of New York. Furthermore he differed from his robber baron peers in supporting labor unions, including the right to strike at will, and their participation in management.

58. Quoted in Samuel Eliot Morrison and Henry Steel Commager, *The Growth of the American Republic*, Oxford University Press, New York, 1937, p. 131.

59. Nevins, *op. cit.*, p. 409. Adams says "he was the more struck by Hewitt's saying, at the end of his laborious career as a legislator, that he left behind him no permanent result except the Act consolidating the Surveys," *op. cit.*, p. 275. But what I'm the more struck by is the contending presence, a hundred and twenty years ago of precisely the same forces—*advancing the same arguments*—that Mack documents for the Landsat satellite.

60. Thompson, *op. cit.*, p. v. The quote stands alone in splendid isolation just after the copyright page as, one presumes, the guiding spirit of the project, not only the Survey's project of mapping America, but Thompson's in writing its "official" description, whose "primary emphasis . . . is on topographic maps" (p. 14).

61. How will it be emptied? By mapping. First the topographers will describe the surface. Then field geologists, using these as base maps, will describe what's underneath. Then the prospectors will show up, mines will sprout, towns, schools, *yellow* school buses. Soon enough it will seem never to have been otherwise. Nevins refer to it as, "the work of reducing the vast trans-Missouri West to the uses of civilization" (*op. cit.*, p. 407). Isn't that perfect? *Reducing it to . . .*

62. In 1992 this optimistic vision is no longer sustainable, because, as Maurice Strong, Secretary General of the United Nations Conference on Environment and Development, recently put it, "The development model which has produced the life styles that we in the industrialized world and the privileged minority in developing countries enjoy is simply not sustainable." He goes on to describe what will be necessary to put us on the path to a more secure and sustainable future: "At the core of this shift there will have to be fundamental changes in our economic life—a more careful and caring use of the earth's resources and greater cooperation and equity in sharing the benefits as well as the risks of our technological civilization" (quoted by Alan P. Ternes, "Great Expectations," *Natural History*, June, 1992, p. 6). The *best* face we can put on the mission of the Survey is no longer a very positive one.

63. These recall to us the rest of Thompson's list of purposes topographic survey sheets might serve: exploring, selecting damsites, locating communication facilities, selecting industrial sites, routing pipelines, planning highways.

64. Rarely has this voice seemed as hollow as when David Love speaks it in John McPhee's *Rising From the Plains* (Farrar, Strauss, Giroux, New York, 1986). Love is a preeminent field geologist of the United States Geological Survey, "the grand old man of Rocky Mountain geology." Born and raised among them, he loves the Wyoming Rockies. When his science leads him to the discovery of oil under Yellowstone National Park, he does not hesitate to follow. McPhee says that, "In pursuing this project, the environmentalist in him balked, the user of resources preferred the resources somewhere else, but the scientist rode on with the rod. He knew he would bring scorn upon himself, but he was not about to stifle his science for anybody's beliefs or opinions"(p. 205). McPhee quotes Love saying, "A scientist, as a scientist, does not determine what should be the public policy in terms of exploration for oil and gas" (p. 204–205), especially when the agency of the government he works for is dedicated to the description of the

resources of the country for *their economic exploitation*. McPhee describes Love's tone of voice as seeming "to exclude both emotion and opinion" (p. 204), that is, as . . . *the voice of science* . . . but one reduced to a simple instrument of the capital which sustains it.

Chapter Five

1. As will become more apparent below, it is not irrelevant that were our legend a photograph in the *National Geographic Magazine*, it is this pendent sentence that would be called the "legend." At the Geographic, caption writing is an art practiced by those in the Legends Division.

2. Arthur Robinson *et al.*, *Elements of Cartography, Fifth Edition*, John Wiley & Sons, New York, 1984, p. 159. It is instructive that, despite their indispensability, legends are granted but two paragraphs in the chapter on *design*, where they play the role of illustrations of the principles of figure-ground relationships. In light of the discussion, below, of the "naturalization" function of myth, it is not surprising that Robinson *et al.* should have said, 'naturally indispensable.'

3. *Ibid.*

4. Ulla Ehrensvärd says, "the role color plays on maps has yet to receive thorough historical scrutiny," and this remains true despite her, "Color in Cartography: A Historical Survey," in David Woodward, editor, *Art and Cartography*, University of Chicago Press, Chicago, 1987, pp. 123–146. See my review in *Cartographica*, *24*(3), Autumn, 1987, pp. 76–82, especially, on color, pp. 80–82.

5. Of course the contradictions here are . . . *terrifying*. Animals and roads don't, after all . . . *mix*. In this afternoon's mail, comes this from James Berry:

> "The rabbits are all gone," someone said. "I haven't seen a rabbit in years; they used to be everywhere." And in Halifax [North Carolina] the other day at a meeting of retired school teachers someone said, "Do you ever see rabbits anymore?" And everybody shook their heads and wondered. And on the way from Raleigh to Chapel Hill Tuesday, I saw six run-over possums and two raccoons and three thousand pushed-over trees and fifty earth movers smoking and chugging and doing the only thing they can do: clearing and grading. So the creatures had to flee. Where could they go? Someone spoke up. "That's what it means to have a job. You have to have a job to get money, and you have to have money to live, and having a job means you have to be doing something, and everything you do changes the world. So you see, it's just the way it is. The creatures have to go. Rabbits and possums and raccoons and trees and woodpeckers and all, what do they matter? Roads! That's what North Carolina's all about. North Carolina's about roads and more roads. And it's about automobiles. You got to be able to go from anywhere to anywhere at sixty miles an hour; without stopping. The creatures can just get out of the way." (James Berry, "It's People or

Rabbits, Reprise, March, 1985," *The Center for Reflection on the Second Law, Circular 146*, May, 1992, p. 1)

And of course . . . this *is* the North Carolina of the road map!

6. This is no longer, if it ever was, quite true, though with 77,058 miles to Texas' 77,075 miles, it's as close as possible (according to "Officials say bridges still get less attention," *News and Observer*, May, 18, 1992, p. B2).

7. Roland Barthes, *Mythologies*, Hill and Wang, New York, 1972, p. 109. Felicitously translated by Annette Lavers, *Mythologies* consists of a number of 'mythologies' followed by the long essay, "Myth Today." It is from this latter that this reference and the following quotation come.

8. *Ibid.*, p. 115–116.

9. *Ibid.*, p. 115.

10. *Ibid., p. 131.*

11. This is even more obvious at the county level: it would be genuinely helpful to distinguish counties prohibiting the sale of alcoholic beverages from those selling beer and wine and mixed drinks. But in fact the carefully delineated counties are not distinguished in any way. Then why show them? It is not a question that can be answered at the level of language. Only on the level of myth is their presence explicable, where North Carolina (and any other state), defender of states' rights (as it has to be), can be seen to dissolve in turn into its constituent counties, their boundaries an unscreened application of the yellow used to demarcate the sovereignties surrounding North Carolina, leaking, as it were, into the state via these county edges.

12. This is the sole acknowledgment of the presence of native Americans in North Carolina, though North Carolina has the largest number of them of any state east of the Mississippi. Is this information that properly belongs on a state highway map? Maybe, maybe not, but at this point it has become difficult to ignore the fact that North Carolina exists at all only because the native Americans were dispossessed of the territory our map so convincingly *possesses* in the name of North Carolina. Brian Harley treats the theme pretty generally in "Victims of a Map: New England Cartography and the Native Americans," paper read at the Land of Norumbega Conference, Portland Maine, December, 1988.

13. In Chapter One we saw how this issue reduced the editors of *The Times Atlas of the World* to gobbledygook. The question is whether mapmakers are ever going to be willing to accept their personal responsibility for the decisions they make, or will forever . . . pass these off onto the world.

14. Or even the fact, highly relevant to motorists, that along with its award for most miles in a state maintained highway system (or close to it), North Carolina *also* gets the award for *most substandard state-owned bridges*. According to *Better Roads*, a transportation trade magazine, 8,286 of the state's 16,828 bridges were either substandard or functionally obsolete. Commenting on the issue, Bill Holman, an environmental lobbyist, observed that part of the trouble is that businesses are more interested in new roads than in improving old ones: "You don't open up new areas to development when you replace a bridge," Holman said ("Officials say bridges still get less attention," *News and Observer*, May, 18, 1992, p. B2). As I write this Barry Yeoman, writing in the Raleigh-area *The Independent*,

has inaugurated a five-part series, "Highway Robbery: How Campaign Dollars Rule the Roads," in the first part of which he documents the relationship between routes and campaign contributions (Barry Yeoman, "Paving Under the Influence," *The Independent*, *10*(21), May 20–26, pp. 8–13). It just underlines our contention—here, in this immediate, local context—that what gets mapped is what makes money for those who have money. And all the rest of it is a kind of technical handwaving.

15. It is also a sixth as many as the state printed of its *1988–1989 North Carolina Coastal Boating Guide* (100,000 copies) and a third as many as it printed of its *North Carolina Variety Vacationland 1989–1990 Aeronautical Chart* (40,000 copies). The state's priorities could not be clearer: road maps, 1.6 million copies; boat maps, 100,000 copies; maps for private planes, 40,000 copies; maps for public transportation, 15,000 copies. North Carolina publishes the edition size and cost per copy on all public documents. Our copy of the *Public Transportation Guide*—the map's second edition—carries a 1985 date. Curiously, although the governor's wife's photograph graces the highway map, it is missing from the public transportation guide, where he stands alone.

16. See, for instance, the beautiful treatment of the "Top Hat, White Tie, and Tails" number from Astaire's *Top Hat* in Gerald Mast's *Howard Hawks: Storyteller*, Oxford University Press, New York, 1982, pp. 21–24, which considers each of these elements (except for Ginger, of course).

17. Umberto Eco, *A Theory of Semiotics*, Indiana, Bloomington. 1976, pp. 48–49.

18. *Ibid.*, p. 49.

19. Jonathan Culler, *The Pursuit of Signs: Semiotics, Literature, Deconstruction*, Cornell, Ithaca. 1981, p. 24.

20. Roland Barthes, *Camera Lucida*, Hill and Wang, New York, 1981, pp. 100–102.

21. These examples come from the verso of "Central America," published as a supplement to the *National Geographic*, April 1986, 466A.

22. The Central America map is as cited above. That of the Central Plains comes from the verso of "Central Plains," published as a supplement to the *National Geographic*, September 1985, 352A.

23. The reference is to the original edition of *The Nuclear War Atlas*, a two-by-four foot sheet with 28 two-color maps recto—in inflammatory black and red—and text verso published by The Society for Human Exploration, Victoriaville, Quebec, 1982; although the Backwell version we have cited previously is socially conscious enough (William Bunge, *Nuclear War Atlas*, Basil Blackwell, Oxford, 1988).

24. Michael Kidron and Ronald Segal, *The State of the World Atlas*, Simon and Schuster, New York, 1981. This was followed by a second edition, *The New State of the World Atlas*, Simon and Schuster, New York, 1984; a third edition, *The New State of the World Atlas Revised and Updated*, Simon and Schuster, New York, 1987. A fourth edition has since been published. It has spawned a whole family of similarly engaged atlases: Michael Kidron and Dan Smith's *The War Atlas*, Pan Books, London, 1983; their *The New Atlas of War and Peace*, Simon and Schuster, New York, 1991; Joni Seager and Ann Olson's *Women in the World*

Atlas, Simon and Schuster, New York, 1986; and Joni Seager's *The State of the Earth Atlas*, Simon and Schuster, New York, 1990. In each of these the violation not only of good cartographic taste, but map reticence about its interests signals . . . *righteous indignation*.

25. See Mark Monmonier's trenchant treatment of the Love Canal issue in *How to Lie with Maps*, University of Chicago Press, 1991, pp. 121–122. With respect to the absence of this infamous toxic waste site on recent Survey quads he argues, "Although both federal and state mapping agencies might contend that topographic maps should only show standardized sets of readily visible, more-or-less permanent features, such assertions seem hypocritical when these agencies' maps routinely include boundary lines, drive-in movie theaters, and other elements far less important to human health." Why couldn't he be equally perspicacious with respect to maps in general? Brian Harley, of course, notes that "Official map-making agencies, usually under the cloak of 'national security,' have been traditionally reticent about publishing details about what rules govern the information they exclude especially where this involves military installations or other politically sensitive sites," in J. B. Harley, "Maps, Knowledge, and Power," in Denis Cosgrove and Stephen Daniels, editors, *The Iconography of Landscape*, Cambridge University Press, Cambridge, 1988, p. 306.

26. Roland Barthes, "The Plates of the Encyclopedia," in *New Critical Essays*, Hill and Wang, New York, 1980, p. 27.

27. The *New York Picture Map* was created by Hermann Bollmann for Pictorial Maps Incorporated, New York. The recto carries Bollmann's rendering of midtown Manhattan in five colors, and the verso a two-color planimetric map of the city of New York. Approximately 34 by 43 inches, the map sheet folds to fit a jacket that includes 48 pages of text. It is not dated. For another approach to a not dissimilar issue, see Edward Tufte's treatment of Constantine Anderson's highly similar axonometric of a nearly identical portion of midtown Manhattan (*Envisioning Information*, Graphics Press, Cheshire, Connecticut, 1990, p. 37). Tufte's conclusion? A most unconventional design strategy: "*to clarify, add detail*."

28. R. L. Gregory, in *Eye and Brain: The Psychology of Seeing* (McGraw-Hill, New York and Toronto, second edition, 1973, pp. 160–176), identifies *personal experience* and the *geometry of environment* as key ingredients of our ability to decode perspective transcriptions.

29. Nikhil Bhattacharya, "A picture and a thousand words," in *Semiotica*, 52(3/4), 1984, pp. 213–246. This, and several of the references that follow, are from this special issue titled *The Semiotics of the Visual: On Defining the Field*, edited by Mihai Nadin.

30. *Pretense* because unlike the *Earth at Night* (W. T. Sullivan, *Earth at Night* Hansen Planetarium, Salt Lake City, 1986), this map is really a map of population distribution, not night lights: *Map GE-70, No. 1, Population Distribution, Urban and Rural in the United States: 1970 (nighttime view)*, Bureau of the Census, U.S. Department of Commerce, Washington, D.C.

31. The distinction being drawn here is essentially the same as that of Hansgeorg Schlichtmann, "Characteristic Traits of the Semiotic System 'Map Symbolism,'" in *The Cartographic Journal*, 22(1), June 1985, pp. 23–30. Schlichtmann differentiates "plan information" from "plan-free information" on the basis

of the former's inclusion of location, and content items contingent thereon (*i.e.*, transcribed shape and extent).

32. Compare, for example, the satellite image reproduced on pages 28 and 29 of the *Atlas of North America: Space Age Portrait of a Continent*, National Geographic Society, Washington, D.C., 1985; or that on page 54 of Michael and Susan Southworth, *Maps: A Visual Survey and Design Guide*, Little, Brown, and Co., Boston, 1982; or, of course, the Hansen map, *op. cit.*

33. The term "metaphor" is used here in the most general sense of representation through a surrogate interpretant. Bethany Johns, in "Visual Metaphor: Lost and Found" (*Semiotica*, 52(3/4), 1984, pp. 291–333), distinguishes between metonymy (whole-for-part metaphor) and synechdoche (part-for-whole metaphor). Some authors invert this terminology. Within written language, distinctions among metaphoric types are numerous; but their applications to graphic signs are largely unexplored and of questionable utility.

34. Barbara S. Bartz, "Type Variation and the Problem of Cartographic Type Legibility—Part One," in *The Journal of Typographic Research*, 3(2), April 1969, pp. 130–135, summarizes the iconic ("analogous") characteristics of letterforms in the cartographic context as those referring to location (point location, linear and areal extent, shape and orientation of feature), quality, quantity, and value (relative importance).

35. Southworth and Southworth, *op. cit.*, p. 189, reproduce two examples; Kevin Lynch reproduces another (*Managing the Sense of a Region*, MIT Press, Cambridge, 1976, pp. 158–159 and dust jacket).

36. Paschal C. Viglionese, "The Inner Functioning of Words: Inconicity in Poetic Language," in *Visible Language*, 19(3), 1985, pp. 373–386, foregrounds these potentials in a series of analyses attentive to the pre-phonographic origins of linguistic expression and the cultural bases of iconicity.

37. In Chapter Three we referred to this by its more familiar name, *projection*, though we actually treated it, explicitly, as a code. By reducing *all* aspects of map production equally to codes, we hope to reveal the similarity among what are usually entirely segregated. Thus, ordinarily, projections are treated as problems in . . . mathematics, but map layouts as ones of . . . design (whence a lot of the old science/art distinction, despite the fact that science can hardly be reduced to math, or art to design). In fact, both are equally . . . *coded* (only the codes are different).

38. A classical example would be the 23 small multiples of Los Angeles air pollution showing the average hourly distribution of reactive hydrocarbons that Tufte illustrates in *The Visual Display of Quantitative Information*, *op. cit.*, p. 170; but Stephen Hall illustrates images he calls maps of phenomenon transpiring in small parts of nanoseconds. See the image of the creation of the first Z particle observed in Stephen Hall, *Mapping the Next Millennium: The Discovery of New Geographies*, Random House, New York, 1992, between pp. 240 and 241.

39. These examples are from J. B. Post, *An Atlas of Fantasy*, Mirage Press, Baltimore, 1973. A revised edition is published by Ballantine Books, New York, 1979.

40. We refer here to the maps occupying pp. 80–81 and 148–149 of *Goode's World Atlas, Sixteenth Edition*, Rand McNally and Co., Chicago, 1982.

41. One might reflect here on the currency of data drawn from geographic

information systems, the difference in time between their point of acquisition and point of use, and the liability potentially incurred. Given the naive tendency of most users to accept any electronically-coded information as current, the onus is clearly on the purveyor of information to inform the user to the contrary. Political bubble-bursting notwithstanding, this is a responsibility that the system manager ignores at his own peril: unearthing a telephone cable is one thing; cracking open an oil tanker is quite another.

42. Recently this similarity has been increasingly acknowledged. See, for example, Nina Siu-Ngan Lam and Dale A. Quattrochi, "On the Issues of Scale, Resolution, and Fractal Analysis in the Mapping Sciences," *Professional Geographer*, 44(1), 1992, pp. 88–98, where "scale" and "resolution" refer equally to spatial, temporal and "spatio-temporal" domains. Note the up-to-date use of "mapping sciences". What Lam and Quattrochi really make clear, however, are the number of *new* avenues for political activity in the process of mapping.

43. Tommy Carlstein, *Time Resources, Society and Ecology*, George Allen and Unwin, London, 1982, pp. 38–64, argues convincingly for a 'time-space' framework of geographic notation. So does Allan Pred, most comprehensively in *Making Histories and Constructing Human Geographies: The Local Transformation of Practice, Power Relations, and Consciousness*, Westview, Boulder, 1990.

44. This map is reproduced, with some fanfare, in Edward R. Tufte, *The Visual Display of Quantitative Information*, Graphics Press, Connecticut, Cheshire, 1983, pp. 41 and 176.

45. The example at hand concludes the *North American Road Atlas* published by the American Automobile Association, Falls Church, Virginia, 1984.

46. *The World Geo-Graphic Atlas: A Composite of Man's Environment*, edited and designed by Herbert Bayer, was produced in 1953 for the Container Corporation of America. Described in the foreword as "an effort to contribute modestly to the realms of education and good taste," it is, as a gesture of corporate good will or a device of corporate promotion (take your pick), an exceptionally lavish and ambitious volume. On the role of "exchange value" at the expense of "use value" in Bayer's involvement with the Container Corporation of America, see Folke Nyberg's comments in his "From *Baukunst* to Bauhaus," *Journal of Architectural Education*, 45(3), May, 1992, p. 136.

47. Which is pretty much, but not quite the story. In his preface to the Blackwell edition, Bunge has this to say about the original, poster version:

> On a brief visit back to Toronto, James Cameron, a geographer at York University, suggested that I do an atlas on nuclear war. York provided newspaper clippings and some cartographic work through the efforts of Gerry Bessenbrugge but soon broke off its involvement. Yet my wife and I persisted, and this resulted in the poster edition of this atlas which was on the streets in June, 1982, just one week too late for the great United Nations demonstration in New York City. The first edition of the atlas was designed for field use among the unemployed of Detroit's black slum ghetto . . . The original edition was in the tradition of *Lobeck's Physiographic Diagram of North America*, with 20,000 words of text on one

side and 28 maps on the other, suitable for poster display upon completion of reading it. The 20 in. x 34 in. poster folded into a 5 in. x 8 in. size designed for peace demonstrations, where it was abundantly sold. Selling the atlas as an excuse to talk peace during the summers of 1982 and 1983, talking to thousands of people door-to-door, often at great length, especially in Toronto, retaught me Detroit's lesson that people needed, as well as a dire warning, hope and a more articulated plan for saving children (William Bunge, *The Nuclear War Atlas*, Basil Blackwell, Oxford, 1988, pp. xxi–xxii).

Although hardly likely to inspire envy among many professional cartographers, this atlas in its poster form assumed the form appropriate to its purpose. It would be hard to imagine as an expensive coffee-table book like the *World Geo-Graphic Atlas* except, perhaps, as a device of the blackest humor.

48. This term is more widely accepted among graphic designers than among linguists. Thomas Ockerse and Hans Van Dijk, "Semiotics and Graphic Design Education," [*Visible Language*, 8(4), 1979, p. 363] describe the supersign as, "a sign which allows for a complex simultaneity of possible interpretants." In "De-Sign/Super-Sign" [*Semiotica*, 52(3/4), 1984, pp. 251–252], Ockerse elaborates on,

> The problem of defining the so-called 'super-sign.' This means to provide a rational system for communication wherein the sum forms the major mode of signification. The participating elements within this complex whole contribute bits of information. The whole is actually a sign made up of other signs; more precisely, the supersign is a sign system. This system is intended to include all signs that operate within the system or that can/will influence the system: the bits, their structural relations, the sum representations created by the juxtapositions of micro- and macro-elements (bits to bits, bits to groups, groups to groups, groups to the whole, the whole to others, etc.). Involved are potential layers and levels of information (in terms of importance, denotative and connotative references) for reader/viewer. The supersign is like a text; but its potential is even intertextual, a characteristic of signs. In fact, the supersign concept even provides a system that invites the reader/viewer to become an active participant in a generative process.

It will become apparent that, in our analysis, the term "system" has a more specific meaning than that intended by Ockerse; but this does not indicate disagreement over the nature or function of the supersign.

49. C. Grant Head, "The Map as Natural Language: A Paradigm for Understanding" (*Cartographica*, *21*(1), 1984, pp. 1–32) stresses two levels of interpretation, citing the following: Barbara Bartz Petchenik, "From Place to Space: The Psychological Achievement of Thematic Mapping," *The American Cartographer*, 6, 1979, pp. 5–12; Judy M. Olson, "A co-ordinated approach to map communication improvement," *American Cartographer*, 3, 1976, pp. 151–159; and Jacques Bertin, "La test de base de la graphique," *Bulletin du Comitrancais de Cartographie*, 79, 1979, pp. 3–18. Among these, however, it turns

out that only Petchenik's analysis is entirely restricted to two levels ("being-in-place" and "knowing-about-space"): Olson's "Level One" and "Level Two" are supplemented by a "Level Three" that is curiously distinct in its attention to meanings; and Bertin, in *Semiology of Graphics* (University of Wisconsin Press, Madison, 1983, pp. 141 and 151), acknowledges a variety of "intermediate" levels between the "elementary" and the "overall". Schlichtmann (*op. cit.*, pp. 25 and 27–28) identifies three levels of signification—"minimal signs, macrosigns, and texts"—which seem to differ more in extent than degree of synthesis. While none of these analyses recognizes a presentational, or discursive, level of signification, our terms are probably in closest agreement with Schlichtmann's.

50. Our concern here is not the neurological processing of stimuli, but the *interpretation* of visual signs. The map user, regardless of—and oblivious to—physiological means, is obviously capable of both composing and decomposing complex signs; one of these abilities is of little use without the other. There seems to be a tendency among cartographers to regard perception as an exclusively constructive—even additive—process, encouraged perhaps by an affinity for mechanistic perceptual models that, for the most part, simply invert the biological metaphors of technological design (offering cameras for eyes, telecommunications systems for neural systems, or industrial robot vision for human cognition), and driven by a virtual obsession with the measurement of responses to largely decontextualized cartographic expressions. But the issue at hand is one of interpretive strategy: a strategy that operates on the organization of meanings, and the construction and deconstruction of *meaningful structures*. Its application is bidirectional and comprehensive.

51. This subject is given thorough treatment by Jacques Bertin, *op. cit.*, pp. 195–268 and 321–408.

52. Paul Klee, *Pedagogical Sketchbook*, Faber and Faber, London, 1968, pp. 18–21. First published in 1925, and first translated in 1953, this, together with Wassily Kandinsky, *Point and Line to Plane* (Dover, New York, 1979), root the Formalist approach to visual design firmly in the curriculum and practice of the Bauhaus. Contemporary treatments of a general nature include Donis A. Dondis, *A Primer of Visual Literacy*, (M.I.T. Press, Cambridge, 1973), Wucius Wong, *Principles of Two-Dimensional Design* (Van Nostrand Reinhold, New York, 1972) and, despite its title, Jacques Bertin's *Semiology of Graphics* (*op. cit.*). For decades, Formalism has dominated the methodology of cartographic design: its appearance in the modern textbook is effectively compulsory, and a bibliography of papers that construct "design guides" from Formalist principles would be too extensive to present here. For a relatively concise, cartographically-oriented, review see Howard T. Fisher, *Mapping-Information: The Graphic Display of Quantitative Information*, Abt Associates, Cambridge, 1982, pp. 60–115.

53. Though why not? The roads on the North Carolina road map are. What, of course, we understand in this way is that "roads" *per se* are not features. Rather federal roads are, state roads are, county roads are, and so on.

54. J. S. Keates, *Understanding Maps*, Longman Group Ltd., London and New York, 1982, p. 82.

55. However the blue line, in and of itself, does represent a road on the North Carolina highway map.

56. Maurice Merleau-Ponty, *Signs*, Northwestern University Press, Illinois, Evanston, 1964, p. 39.

57. In the case of cadastral maps this other sign system is often purely linguistic (the description of the boundary, the names of the owners, and so on).

58. This term is used in the sense intended by Peirce: to express a causal relation between object (steep slope, river, city) and interpretant (twisting road, parallel roads, circular highway segment). For Peirce, *icon*, *index* and *symbol* constitute the second of three trichotomies which jointly define and elaborate taxonomy of signs. See Charles Sanders Peirce, *Philosophical Writings of Peirce*, Dover, New York, 1955, pp. 98–119, or *Collected Papers of Charles Sanders Peirce, Vol. II, Elements of Logic*, Harvard University Press, Cambridge, 1960, pp. 134–173.

59. The familiar example of the musical theme, which retains its identity despite transposition to another key or rescoring for a different ensemble of instruments, is remarkably evocative of the cartographic sign system that retains *its* identity throughout numerous topological and scalar transformations, spatial reorientations, and symbolic representations. Clearly, the recognizable whole, in both cases, is an artifact of structure rather than sensation—a *gestalt*.

60. Bill Bunge made a similar point with his map "The Continents and Islands of Mankind," which shows—against a white ground in black—simply those portions of the globe harboring more than 30 persons per square mile. Period. About the map he made these comments:

> When the original explorers went out they searched for people too, for instance, good slaves. But mapping people was very dangerous. People are also mobile. Compared to mountains, rivers, coastlines, they are nearly invisible. But at least the names of 'tribes' were placed on original maps. And as this material was accumulated it became known as 'the map'. It became the stuff of the 'base map'. And once the 'base map' for a region was complete, it was 'explored'. It has been impossible evidently to conceive even philosophically of a more appropriate base map for our times. We use as the absolute irreducible element the distinction between what is wet and what is dry. Might it not be better to distinguish between what is populated and what is empty of people? The deserts of the world, the ice caps, have more in common with most of the oceans than with South Asia. The North Atlantic, with its permanently transient population, might be better classified with Iowa than the South Pacific. Even recognizing that some human interest has always been shown in humans—the priorities have been so reversed that the base map itself should be reexamined. It might be sanguine to start having grade school children around the world memorizing the continent and islands of people as the basic ingredient in their mental maps. (W. Bunge, *Detroit Geographical Expedition, Field Notes*, *1*, 1969, p. 2).

61. Kidron and Segal, *op. cit.* This atlas presents 57 map plates, and corresponding micro-essays, addressing urgent (and frequently controversial) socio-political issues of global scope. Its overcrowded page layouts, animated

symbolism, disturbing colors, pointed titles, and terse text form the ingredients of an acerbic discourse on the corruption and repression of the modern nation–state.

62. Robert Scholes, *Semiotics and Interpretation*, New Haven: Yale University Press, 1982, p. 144.

63. *Ibid.*, p. 34.

Chapter Six

1. In order: A. A. Milne, *Winnie-the-Pooh*, Dutton, New York, 1926. The map was drawn by Ernest H. Shepard no matter what it says. J. R. R. Tolkien, *The Hobbit*, Houghton Mifflin, Boston, 1938. The maps were drawn by the author's son, Christopher Tolkien. Arthur Ransome, *Swallows and Amazons*, Jonathan Cape, London, 1931. The author drew the maps. T. H. White, *Mistress Masham's Repose*, Putnam's, New York, 1946. The endpapers were drawn by Raymond McGrath. Fritz Muhlenweg, *Big Tiger and Christian*, Pantheon, New York, 1952. The map itself is uncredited. Robert Louis Stevenson, *Treasure Island*, Scribner's, New York, 1981. This is a reissue to commemorate the 100th anniversary of Stevenson's tale. The map is uncredited, but surely was not drawn by N. C. Wyeth. Goscinny and Uderzo, *Astérix le Gaulois*, Darguad, Paris, 1961. Uderzo drew the map. Holling C. Holling, *Paddle-to-the-Sea*, Houghton Mifflin, Boston, 1941. All of the great Holling books have maps in them, including early ones like *The Book of Indians* (Platt and Munk, New York, 1935). Gertrude Crampton, *Scuffy the Tugboat*, Simon and Schuster, New York, 1946. The book was illustrated and the map drawn by Tibor Gergely.

2. This point will be implicit in what follows, but since this is true, if teachers would only relax and treat maps like English, they would find that kids can learn it like English. I am not claiming here that humans posses a Map Acquisition Device like the Language Acquisition Device hypothesized by Noam Chomsky, but I am claiming that maps terrify teachers more than students. Kids who can't understand them in school, come home to draw them of treasure buried in the backyard. I agree, therefore, with Henry Castner (*Seeking New Horizons: A Perceptual Approach to Geographic Education*, McGill-Queen's University Press, Montreal, 1990) when he argues that map learning should not begin with "mastery of conventional symbols of maps" (p. 96) and when he insists that students should learn *with* maps rather than *about* them. But I don't think the special treatment he recommends (by analogy with the music education strategies of Suzuki and Orff) is necessary. If taken for granted by the teacher, they would be taken for granted by the students . . . because they are a taken-for-granted aspect of our lives.

3. I'm looking at the plate in Stephen Hall's *Mapping the Next Millennium* (Random House, New York, 1992), between pp. 240 and 241. Not all maps of this phenomenon are as abstract.

4. Or computer screen. Anyone who thinks any of this will change when the map leaves the paper for the "virtual environment" hasn't understood a thing. Return to Chapter One: *start over*.

5. Norman Thrower, *Maps and Man*, Prentice-Hall, Englewood-Cliffs, 1972, p. 78.

6. Edward Lynam, *The Map Maker's Art*, The Batchworth Press, London, 1953, p. 38.

7. Arthur Robinson, *et al.*, *Elements of Cartography, Fifth Edition*, John Wiley & Sons, New York, 1984, p. 368. These statements are less strong than those in earlier editions, where landform relief was regarded as "so different from all the others as to make it almost imperative that it be treated separately" (Arthur Robinson and Randall Sale, *Elements of Cartography, Third Edition*, John Wiley & Sons, New York, 1969, p. 172).

8. David Greenhood, *Mapping*, The University of Chicago Press, 1964, pp. 74–75. This is precisely the ingenuity required to envision information in general according to Edward Tufte: "All communication between the readers of an image and the makers of an image must now take place on a two-dimensional surface. *Escaping this flatland is the essential task of envisioning information—for all the interesting worlds (physical, biological, imaginary, human) that we seek to understand are inevitably and happily multivariate in nature. Not flatlands.*" (*Envisioning Information*, Graphics Press, Cheshire, Connecticut, 1990, p. 12). Making the problem more general does not make it less daunting.

9. See the sharply abbreviated "Historical Background" in Robinson *et al.*, *op. cit.*, pp. 369–372; as well as the more convincing treatment by Eduard Imhof in his, *Cartographic Relief Representation*, de Gruyter, Berlin, 1982, pp. 1–14.

10. Again, this reflects the distinction drawn in Chapter Two between *mapping* and *mapmaking* or *map-immersed* societies.

11. Leo Bagrow, *History of Cartography*, Harvard University Press, Cambridge, 1966, p. 27. This is a history, once regarded as definitive, which has *not* aged well.

12. John Spink and D. W. Moodie, *Eskimo Maps from the Canadian Eastern Arctic: Cartographica Monograph No. 5*, 1972, p. 16.

13. *Ibid.*, pp. 16–17.

14. *Ibid.*, pp. 16–17.

15. As discussed in Rudy Wiebe, *Playing Dead*, NeWest, Edmonton, 1989. Wiebe quotes Gagné to this effect: "All visible phenomena . . .—whether things, beings, places, areas, or surfaces—are viewed two-dimensionally." This introduces a rather brilliant analysis of intercultural contact, but I am not at all certain what to make of it.

16. Robert Rundstrom, "A Cultural Interpretation of Inuit Map Accuracy," *Geographical Review*, 80(2), April, 1990, p. 165. Rundstrom stresses the relationship between Inuit mimicry and mapping.

17. Okay, this is pushing it. But the fact is, until we have a reasonable anthropology of map use in our culture this is about as good as the evidence gets. The terrifying fact is that we know much more about Inuit map use than we do about contemporary American map use.

18. Rundstrom, *op. cit.*, p. 163.

19. *Ibid.*, p. 166.

20. This would hardly be unprecedented. In his wonderful *The History of Topographical Maps: Symbols, Pictures and Surveys* (Thames and Hudson,

London, 1980), P. D. A. Harvey documents a similar situation with respect to large-scale mapmaking in medieval Europe: "It is quite possible that many fifteenth-century maps were drawn by people with no previous knowledge of maps at all, who saw themselves as doing something entirely new, using a quite original method to set out landscape on paper or parchment" (p. 89). Because they were isolated, no tradition developed, that is, no semiological tradition (no store of signs developed on which other mapmakers could draw: all had to be invented, or borrowed from other graphic traditions).

21. *Ibid.*, pp. 37–42.

22. For the Luba, see Thomas Reefe, "Lukasa: A Luba Memory Device," *African Arts, 10*(4), 1977, pp. 48–50, 88; and Mary (Polly) H. Nooter, "Secret Signs in Luba Sculptural Narrative: A Discourse on Power," in Christopher Roy, editor, *Art and Initiation in Zaire: Iowa Studies in African Art, 3*, 1990, pp. 35–60.

23. This is sarcastic. As a complete child of his time and place, Bagrow regarded all non-Europeans as "primitive." This not only underwrote his treatment of such cartographic products as he was able to unearth, but his willingness as well to imagine that even (evidently) great, high cultures such as that of India might be devoid of a mapmaking tradition. Here is Joseph Schwartzberg on the situation: "Until quite recently, conventional wisdom among historians of cartography was that India lacked any indigenous cartographic tradition. This bit of received knowledge was encapsulated in two statements in Bagrow's once canonical *Die Geschichte der Kartographie*, which, as translated by Skelton, read, 'India had no cartography.' When I agreed in 1979 to write on indigenous Indian cartography, in response to a request from Harley and Woodward, editors of *The History of Cartography*, I had no reason to doubt the correctness of Bagrow's judgment and wondered at the time whether I could even write so much as the suggested 3,000 words on the subject. Today, thirteen years and about 175,000 words later, I realize that I was mistaken. So too were Bagrow and many others" ("An Eighteenth Century Cosmographic Globe from India," paper presented at the 88th Annual Meeting of the Association of American Geographers, San Diego, April, 1992, p. 1). The almost universal willingness to subscribe to Bagrow's essentially idiographic depiction of world cartographic history reflected the complete absence of *any* theoretic understanding about the relationship of mapmaking to other social and historical forces.

24. Bagrow, *op. cit.*, p. 27, does note that the ancient cultures of Mexico "were highly developed" but this is just after having remarked that "many savage peoples have shown some skill in drawing maps." The linkage, however unforgivable, was again more a product of Bagrow's times than of his personality. Thrower's remark is on page 10 of his *Maps and Man, op. cit.*

25. "Essentially, however, the pictorial conventions of Zacatepec 1 are identical with the conventions of pre-Conquest manuscripts discussed in Chapter III. Dates, names of persons, marriage and conference, roads and warpaths, conquest and place signs are for the most part represented in their pre-Conquest idiom," Mary Elizabeth Smith, *Picture Writing from Ancient Southern Mexico: Mixtec Place Signs and Maps*, University of Oklahoma Press, Norman, 1973, p. 92 *et passim*. See also Donald Robertson, *Mexican Manuscript Painting of the Early Colonial Period: The Metropolitan Schools*, Yale University Press, New Haven, 1959; and Donald Robertson, "The Mixtec Religious

Manuscripts," Howard Cline, "Colonial Mazatec Lienzos and Communities," and Alfonso Caso, "The Lords of Yanhuitlan," all in John Paddock, editor, *Ancient Oaxaca*, Stanford University Press, Stanford, 1966. The Codex Nuttall is available, in color, as *The Codex Nuttall*, Dover Publications, New York, 1975.

26. Smith, *op. cit.*, p. 21.

27. *Ibid.*, p. 39.

28. *Ibid.*, p. 92.

29. But once they did come, Zacatepec *was* connected to them through the map. From 1540, when *Zacatepec 1* was presumably made, until 1892, it remained in Zacatepec. Then it was taken by the citizens of Zacatepec to Mexico City where it was used as corroborating evidence in a land suit.

30. J. B. Harley, "What Happens When We've Made a Map," unpublished lecture given at Pennsylvania State University, 1991, p. 9.

31. Smith, *op. cit.*, p. 94.

32. As this Mixtec system of representation is precedent to the mapping which represents landscape, so it is also precedent to what I. J. Gelb calls writing. Rather, it is what he terms a limited system, a forerunner of writing (*A Study of Writing*, University of Chicago Press, Chicago, 1963, pp. 51–59). The important point is that the Mixtec histories and maps represent . . . *an early system*, in which representational systems later to be teased apart are still fused.

33. An identical drive apparently motivated the development of Mayan record-keeping systems as well, a fact increasingly well-established as the Mayan hieroglyphic code is increasingly well-deciphered. See Linda Schele and Mary Ellen Miller, *The Blood of Kings: Dynasty and Ritual in Maya Art*, Braziller, New York, 1986; and Jeremy Sabloff, *The New Archeology and the Ancient Maya*, Scientific American Library, New York, 1990.

34. Smith, *op. cit.*, pp. 93–96 *et passim*; Bagrow, *op. cit.*, Plate III. The European elements introduced into *Zacatepec 2*—created 50 years later—represent the landscape itself, although the place names retain their logographic form. This means you can find hillsigns representing hills *named* by hillsigns representing *names*. On contemporary maps, hills are coded iconically and their names linguistically. Thus, this sequence: first, hillsigns as names; then hillsigns as hills and names; finally, hillsigns as hills.

35. Bagrow, *op. cit.*, Plate V.

36. Not that they ever separate completely. The linguistic code in Western cartography makes this clear, as does the dominant Oriental tradition in which paintings are written on as a matter of course (see Michael Sullivan, *The Three Perfections: Chinese Painting, Poetry and Calligraphy*, Thames and Hudson, London, 1974) and a minor one in the West (see, among others, no author, *Lettering by Modern Artists*, Museum of Modern Art, New York, 1964; and Norma Ory, *Art and the Alphabet*, Museum of Fine Arts, Houston, 1978.; articles appear on this and related topics frequently in the journals *Visible Language* and *Word and Image*).

37. See Denise Schmandt-Besserat, *Before Writing: From Counting to Cuneiform, Vol. 1*, University of Texas Press, Austin, 1991, for the case in general.

38. It is at this point that conventional histories of hillsigning begin, not only with the lateral or profile view, but with precisely this map. See Imhof, *op. cit.*, p. 1.

39. J. K. Wright, *Geographical Lore in the Time of the Crusades, American Geographical Society, New York, 1925, pp. 252–253.*

40. This is a summary of Lynam, *op. cit.*, pp. 38–41. There was a third parallel development from hand-drawn manuscript maps, through woodcuts, to copper engravings.

41. R. A. Skelton, "Cartography," in C. Singer, E. Holmyard, A. Hall and T. Williams, editors, *A History of Technology: Volume IV: The Industrial Revolution*, Oxford University Press, Oxford, 1958, p. 611.

42. D. H. Fryer, "Cartography and Aids to Navigation," in Singer *et al.*, *op. cit*, *Volume V: The Late Nineteenth Century*, p. 439.

43. Robinson and Sale, *op. cit.*, p. 177.

44. *Ibid.*, p. 173. To gauge the accuracy of their prediction take a look at Thelin and Pike's digital terrain map of the United States, computed from 12 million terrain heights "extracted by semiautomated techniques from the more than 450 1:250,000-scale topographic sheets covering the 48 states" and then machine-shaded. Few recent maps have excited such attention. See Peirce Lewis, "Introducing a Cartographic Masterpiece: A Review of the U. S. Geological Survey's Digital Terrain Map of the United States, by Gail Thelin and Rickard Pike," Richard Pike and Gail Thelin, "Visualizing the United States in Computer Chiaroscuro," and Stuart Allan, "Design and Production Notes for the Raven Map Editions of the U. S. 1:3.5 Million Digital Map," all in *Annals of the Association of American Geographers*, 82(2), June, 1992. The map was lavishly reviewed in *The New York Times*, *Scientific American* and elsewhere.

45. Erwin Raisz, *Principles of Cartography*, McGraw-Hill, New York, 1962, pp. 88–89.

46. My copy bears no indication by or for whom it was created. I presume it was manufactured here for Japanese tourists, but my letters to a number of the retailers noted on the map have borne no fruit.

47. Although probably not representative of anything, the sketch maps discussed were drawn by the following: small town Puerto Rican kids between the ages of 7 and 21; small town Mexican kids between the ages of 13 and 18; junior high students from Massachusetts and North Carolina; high school students from the eastern half of the United States (Massachusetts, Kansas and North Carolina); and adults, both in adult education courses and unassociated with school, from the same places, including a number of secondary school teachers.

48. Frank Klett and David Alpaugh, "Environmental Learning and Large-Scale Environments," in Gary T. Moore and Reginald G. Golledge, editors, *Environmental Knowing*, Dowden, Hutchinson and Ross, Stroudsburg, 1976, pp. 121–131. By "large-scale" the authors mean maps of large areas (that is . . . small scale maps). David Stea has reproduced two other maps from the San Fernando set—in color!—in his *Environmental Mapping: Unit 14, Art and Environment, A Second Level Interdisciplinary Course*, The Open University Press, Milton Keynes, England, 1976.

49. The only results made public have been, Denis Wood, "Early Mound-Building: Some Notes on Kids' Dirt Play," paper presented at the Annual Meeting of the American Association for the Advancement of Science, Boston, February, 1976 (Peter Gould discusses this work in *The Geographer at Work*, Routledge and Kegan Paul, London, 1985, p. 254), and those portions of this

chapter previously published as "Cultured Symbols," *Cartographica*, *21*(4), Winter, 1984, pp. 9–37.

50. Denis Wood, *Fleeting Glimpses: Adolescent Images of that Entity Called San Cristobal las Casas, Chiapas, Mexico*, Clark University Cartographic Laboratory, Clark University, Worcester, Massachusetts, 1971. The number of mappers representing the whole town was 176; their neighborhoods, 92.

51. Denis Wood, "A Neighborhood Is To Hang Around In," *Children's Environments Quarterly*, *1*(4), Winter, 1984–85, pp. 29–35. Typically, the drawing reproduced on page 34 *indicates* but does not *represent* "Vernon Hill."

52. In fact societies in which everyone maps are *mapping societies*. In map*making* societies, not even most of those employed in making maps actually make maps. As a diagram in Morris Thompson's *Maps for Americans* (Department of the Interior, Washington, D.C., 1979, p. 24) makes clear, the process of making maps in advanced mapmaking societies is . . . completely fragmented . . . into planning, aerial photography, control and completion surveys, photogrammetric surveys, cartography (picture of a man bent over a drafting table, cartography reduced to . . . drafting), reproduction and distribution. Patrick McHaffie argues that "as the cartographic production process has been progressively fragmented, the cartographic laborer has become increasingly alienated from the product of his/her labor: cartographic information," and concludes that this "produced increasingly precise and some might say dehumanized maps" ("Restructuring the Public Cartographic Labor Process in America," paper presented at the 88th Annual Meetings of the Association of American Geographers, San Diego, 1992, p. 1).

53. Robinson and Sale, *op. cit.*, p. 172.

54. Nearly half of these drawings were collected by Betty Murrell and her colleagues Greg Wall, Scott Stone and Jeff Schoelkopf, undergraduate students in the School of Design at North Carolina State University at Raleigh. Others were collected by Dick Henry, Nann Boggs, Aileen Kennedy and others, also undergraduates at the same school. The rest were collected by me. All were collected under my supervision.

55. For the ability of kids to produce such drawings see Roger Hart. *Aerial Geography: An Experiment in Elementary Education, Place Perception Research Report No. 6*, Clark University, Worcester, 1971; J. M. Blaut, *Studies in Developmental Geography, Place Perception Research Report No. 1*, Clark University, Worcester, 1969; J. M. Blaut and David Stea, *Place Learning, Place Perception Research Report No. 4*, Clark University, Worcester, 1969; J. M. Blaut and D. Stea, "Studies of Geographic Learning," *Annals of the Association of American Geographers*, *61*(2), 1971, pp. 387–393.

56. I should observe that the hills reproduced here represent *all* the hills Randall drew *as such* in the over 300 drawings he made prior to his fourth birthday. (Though not drawn *as such*, he did introduce slope into his drawings of, for example, railroads; and did call these "hills". In the 13 years since turning four he's drawn . . . *thousands of hills*.)

57. The significance of this will not be lost on those students of the effects of woodcutting or engraving on the history of cartography.

58. In particular those models developed by Heinz Werner and Jean Piaget. For an introduction to the work of the former, see Heinz Werner, *The Comparative*

Psychology of Mental Development, International Universities Press, New York, 1948; for an introduction to the latter, try J. H. Flavell, *The Developmental Psychology of Jean Piaget*, Van Nostrand, New York, 1963.

59. This is to ignore the manifold layers of meaning shoveled onto these fundamental properties. See, for example, the reconstructed Indo-European roots of the words "hill" and "mountain" in *The American Heritage Dictionary* (Houghton Mifflin, Boston, 1969).

60. It should be recalled that the North Carolinians producing the 500 drawings undergirding this analysis came from the mountains, the Piedmont and the coastal plain. Had the physical environment been of much significance in the formation of the hillsign, regional variations should have been manifest in our drawings. None, however, could be observed, though such variation constituted the substance of our strongest hypothesis.

61. My son Randall's experience is illuminating. Although he lives in Raleigh, on the edge of the Piedmont, there are few places from which he could see any kind of slope in profile. In his daily environment—actually situated on the crown of a hill—there were none at all; in his weekly environment only a couple, one of which necessitated his being in a parking lot from which he could catch a profile view of a viaduct. Most of the hills in the environment are associated with ridges of which any distant view reveals only the straight and level horizon of the ridge top. The few hills—some of which are ridge ends separated from the ridge by railroad or other cuts—disappear into this background. There are few enough distant views. Most of these are blocked by buildings and trees. The issue of simple size is also relevant. Danny Amaral reported from Worcester, Massachusetts—a town of "seven hills"—the refusal of his son and a friend, age 9, to accept their location in his home as being on top of a hill until he recounted each of the "hills" one had to climb—in all four directions—to reach it. Its top was "too large" to be on top of a hill. Only in nonurban mountain situations, as a rule, is there frequent and repeated opportunity to view relief in profile, and yet even here—for the young child—the view is blocked more often than an adult will readily accept by trees, homes, cars, and other large structures. The world we take for granted—composed of many years of experience not only of the physical world itself but the movies, postcards and travel posters—is not the world inhabited by the child.

62. For more information on sand play—and its relation to the hillsigning equation—see my "Early Mound-building," *op. cit.*

63. The original edition of *The Little Red Caboose* was published as "A Little Golden Book" by Simon and Schuster, New York, 1953. It contained five pages of illustrations missing from my "A Golden Book" edition published by Golden Press of Racine, Wisconsin, at least missing from the Seventeenth Printing (1976). This edition also contains two pages of illustrations not found in the original edition. You can find the book in all sorts of other formats as well. The following citations are from the original edition, but the information about the number of copies sold appears in the back of the Seventeenth Printing.

64. *Ibid.*, pp. 12–13.

65. *Ibid.*, pp. 16–17.

66. Watty Piper's *The Little Engine That Could*, with illustrations by George and Doris Hauman, has an even more complex publishing history. My edition is

"The Complete, Original Edition Retold by Watty Piper," Platt and Munk, New York, 1961, "from the *Pony Engine* by Mabel Bragg, copyrighted by George H. Doran." This story has been pirated about as frequently as any I can think of.

67. *Ibid.*, pp. 30–36.

68. Wanda Gag, *Millions of Cats*, Coward-McCann, New York, 1928. The book was once also available as part of *Wanda Gag's Story Book*, Coward-McCann, 1954.

69. *Ibid.*, *p.* 3.

70. Virginia Lee Burton, *Choo Choo: The Story of a Little Engine Who Ran Away*, Houghton Mifflin, Boston, 1937, pp. 11–13.

71. Virginia Lee Burton, *The Little House*, Houghton Mifflin, Boston, 1942. In the story, the little house, originally in the country, gets swallowed up by the city, from which it is saved at the end and moved back to the country. The hill it sits on in the country is magically absent in the city. On page 15, we do see a steam shovel (Mike Mulligan and Mary Anne?) cutting through a hill for a road, but the house still sits on its hill on page 17. But on page 19, with the house really in the city for the first time, the hill has entirely vanished. It doesn't reappear until we are far out in the country again with the little house on page 36. The absence of the hill is one of the things, difficult to put your finger on, that makes the city so dreadful.

72. Virginia Lee Burton, *Mike Mulligan and His Steam Shovel*, Houghton Mifflin, Boston, 1939.

73. *Ibid.*, pp. 6–7.

74. *Ibid.*, pp. 16–17.

75. George Wharton James, *Indian Basketry*, Dover, New York, 1972; Herbert Spinden, *A Study in Mayan Art*, Dover, New York, 1975. Both James and Spinden attempt to deal with the simultaneous existence in a culture of "geometric," naturalistic, and conventionalized modes of representation by positing the continuous interaction of (1) the effects of the media of representation, (2) the effects of increasing skill in a given medium, (3) the effects of transferring an image from one medium to another, and (4) the effects of an innate tendency toward naturalistic imitation. (In our case "nature" would be the world presented in the kids' books.) Convergent evolution is James' version of the common evolutionary phenomenon of convergence, but see the originals for examples and subtleties.

76. I am using "spontaneously" in its ordinary sense, as in "spontaneous combustion," not in its technical Piagetian sense.

77. The early history of arithmetic is no clearer than the early history of cartography, but an exciting place to begin is with the papers in Parts II and III of Volume I of James R. Newman, editor, *The World of Mathematics*, Simon and Schuster, New York, 1956. The first 200 or so pages of Morris Kline, *Mathematical Thought from Ancient to Modern Times*, Oxford University Press, New York, 1972, are also stimulating. The topic is of enormous scope, and I apologize for my simplistic treatment of it here.

78. That the one might be taught before the other, however, goes without saying. I recently taught a neighborhood kid to multiply although it turns out he did not know how to add. *That* possibility never occurred to me and since I taught the multiplication by rote, with flash cards—as indeed I learned it—the

issue never arose. The fact that multiplication is employed in long division makes that case substantially different, but in fact the order in which we teach things is not at all independent of the history of their creation, invention or discovery.

79. A. N. Whitehead, *An Introduction to Mathematics*, The Clarendon Press, Oxford, 1911, p. 59. J. M. Hammersley has glossed this passage in his "The Technology of Thought" (in Jerzy Neyman, editor, *The Heritage of Copernicus*, M.I.T. Press, Cambridge, 1974, p. 396) as follows: "The fashionable program of modern mathematics in the high schools errs if it tries to make children conscious of what they are doing when they do arithmetic. It is better to drill arithmetic into the child as second nature. Of course education should teach us how to think; but education is lopsided unless it also teaches us when and how not to think. You cannot ride a bicycle until you have forgotten how you balance." Whatever the ethologic merit of his argument, it is the one that has been used to push arithmetic, reading, writing and hillsigning down the pedagogic and ontogenetic ladders. If you are enamored, as Hammersley is, with computers, you are enamored of complete and unthinking mastery of the multiplication tables as well. The once fashionable program of modern mathematics is, it might be noted, no longer fashionable. Whether or not kids learned the commutative law (and no one has said they didn't), they didn't learn their times tables! But what good is a sheet metal worker who understands the commutative property but can't calculate the area of the shape he's cutting (and so maximize the number of such forms from a given sheet of steel)?

80. It must never be forgotten that structuralists tend to pay attention to the *information* being manipulated by their subjects: children, primitives, psychotics. Though their goal may be to model the *cognitive mechanism* via its significant structures, they only have access to *social behavior*. Their notion of Thought, pristine and biologic, essentially divorced from the social, cultural and ecological context—which happens to underwrite psychology as a distinct field of study—will begin to fade as their work is *subsumed* into higher order syntheses such as that of Humberto Maturana (and Francisco Varela, *The Tree of Knowledge: The Biological Roots of Human Understanding*, New Science Library, Boston, 1988).

81. This runs, of course, completely contrary to the accepted role of the forces modeled by developmental psychologists and cultural anthropologists. I am arguing that child development is very largely a matter of acculturation—not maturation—and would most effectively and realistically be studied, using the existent academic taxonomy, by the anthropologists; whereas adult development is *much* more a matter of development (in the sense usually applied to children) than has been admitted, and that it would be most profitably studied by developmental (child) psychologists. The generation of successfully novel responses to continuous environmental change (embracing the social as well as other environments) smacks far more of what is generally understood by DEVELOPMENT than the iterative maturation processes of growing children. Adults are largely responsible for the ethnogenetic development which is slowly and subsequently inducted into the acculturation mechanism surrounding the child. And while we know a good deal about ethnogenesis taken as a whole, we know next to nothing about the individual development (sometimes called

creativity) which fuels it. Until we can reasonably model this behavior in context we will have understood nothing about development in *any* of its manifestations.

82. Nose, chin, leg, foot, hand: just about everything was originally presented in profile. The issue of this profilism and the birth of the frontal view is gaining increasing attention, and not just from art historians. For a good introduction, start with the first third of E. H. Gombrich's *The Story of Art*, Phaidon Press, London, 1950. Jan B. Deregowski summarizes the problem from the anthropologic perspective succinctly in his "Illusion and Culture," in R. L. Gregory and E. H. Gombrich, editors, *Illusion in Nature and Art*, Scribner's Sons, New York, 1973; Cecelia F. Klein devotes an entire monograph to the rare instances of frontality in Mesoamerican art—and so highlights the significance of the profile view—in her *The Face of the Earth: Frontality in Two-Dimensional Mesoamerican Art*, Garland Publishing, New York, 1976; and Bill Holm refocuses many of the questions in his *Northwest Coast Indian Art*, University of Washington Press, Seattle, 1965.

83. For example, "Mixtec manuscript painting was an art of the ruling class. The pictorial histories were painted by and for the ruling families, to describe events participated in only by persons of rank. In addition, the art of pictographic writing utilized in these manuscripts was practiced by a very few painters and was understood by a few specially trained persons who had memorized the stories and been taught the pictorial conventions used to depict them. According to Padre Burgoa, the painted manuscripts were executed by the sons of noble families who were chosen to be priests and were taught from childhood the art of pictographic writing," and so forth and so on (Smith, *op. cit.*, p. 20). The same was true in every society.

84. The same seems to be the case with children. It is only once they begin to *use* the hills (for bike-riding, skateboarding, in games, as props in drawings featuring other things, and so on) that they begin to present them frontally. This impression is so strong that data are being currently gathered to specifically address this point.

85. But see Francois de Dainville, "De la profondeur a l'altitude," *International Yearbook of Cartography*, 1962; the relevant portions of Harvey, *op. cit.*; and Arthur Robinson, *Early Thematic Mapping in the History of Cartography*, University of Chicago Press, Chicago, 1982, explicitly pp. 210–218, but elsewhere as well.

86. From Calixta Guiterras-Holmes, *Perils of the Soul: The World View of a Tzotzil Indian*, Free Press of Glencoe, New York, 1961, page 217.

87. The interviews were conducted by Dick Henry, at the time an undergraduate student at the School of Design, North Carolina State University, as part of a project on kids' perceptions of the geomorphic environment. I did the transcriptions and editing.

Chapter Seven

1. Patrick Waldberg, *Surrealism*, McGraw-Hill, New York, no date, p. 24. Waldberg's text (pp. 23–24) reads: "In this connection, a surrealist map of the

world drawn in 1929 is extremely significant (*Variétés*, Brussels, June 1929). The only cities shown are Paris and Constantinople, but without France or Turkey. Europe consists only of Germany, Austria-Hungary and an immense Russia, which also takes up half of Asia (the other half of which is composed of China, Tibet and an outsized Afghanistan next to a rather small India). By contrast, the islands of the Pacific occupy two thirds of the world and carry as banners the marvelous names of Hawaii, the Solomons, New Hebrides, New Zealand, the Marquesas and the Bismarck Archipelago. The North American continent, from which the United States is missing, presents a gigantic Alaska, the Charlotte Islands, Labrador and Mexico. Further down, Easter Island is as large as all of South America, which is reduced to one single country: Peru. Childish as this drawing of an imaginary world (considered to be the only desirable one) may be, it corresponds to the permanent orientation of the surrealist ideal. That ideal tends to challenge Western Christian civilization . . . " and so forth and so on. Interest is smeared all over this map, which, ironically, was derived from a Eurocentric Mercator.

2. Mary Elizabeth Smith, *Picture Writing from Ancient Southern Mexico: Mixtec Place Signs and Maps*, University of Oklahoma Press, Norman, 1975, pp. 92 and 288. Smith notes: "This sign consists of a hill and a butterfly, and thus expresses San Vicente's Mixtec name, which is reported by the officials of Zacatepec to be *yucu ticuvua*, or 'hill of the butterfly' " (p. 92).

3. Joby Warrick, "Drug abuse rises in coastal areas: interstates still main corridors," *News and Observer*, May, 12, 1992, p. B1

4. Peter Gould, Joseph Kabel, Wilpen Gorr and Andrew Golub, "AIDS: Predicting the Next Map," *Interfaces*, *21*(3), May-June, 1991, pp. 80–92.

5. This is even more evident in a *Science* story about this work. Under three full color illustrations of the spread of AIDS on the West Coast (1981, 1983, 1988) it says, "Spreading stain: Geographers haven't played a big role in characterizing the AIDS epidemic, but Peter Gould and his colleagues at Pennsylvania State University aim to change that. These maps show the spread of AIDS in the western United States from 1981 to 1988. They are part of a series—one map per year—generated by Gould and his team. In a session at last week's meeting, Gould showed the maps sequentially, saying they depict a 'spatial-geographic logic unfolding over time.' He compared the pattern to a 'wine stain on the tablecloth,' moving from urban centers into the suburbs and surrounding countryside. The epidemic first appeared in Los Angeles and San Francisco, followed by Las Vegas and Phoenix. By 1988 Seattle and Portland had entered the picture and the epidemic continued to intensify through the rest of the decade. In his talk, Gould lashed out at AIDS modelers and government officials for failing to take into account the geographic aspects of the spread of HIV infection. 'Despite millions of dollars spent on the AIDS epidemic,' he said, 'we have virtually no picture . . . [of] the geographic dimensions of this deadly virus. . . . All modeling and forecasting is devoted to a simplistic, and essentially useless, computation of numbers down the time line, ignoring totally the spatial . . . dimensions of human existence.' Gould attributed this lack to 'sheer ignorance, aided and abetted in too many instances by individual and bureaucratic arrogance'" (J. B., *Science 251*, p. 1022). Here the map is elevated into a whip with which to beat those whose ineptitude interferes with progress

in dealing with the epidemic. (For a complementary perspective, see Gary Shannon, Gerald Pyle and Rashid Bashshur, *The Geography of AIDS*, The Guilford Press, New York, 1991.)

6. Jacques Bertin, *Graphics and Graphic Information Processing*, de Gruyter, Berlin, 1981, p. 16. This is increasingly likely to be the case as maps become interactive. "Software for interactive, choroplethic mapping typically focuses on design, with the assumption that the resulting map will be used like a traditional paper one. We believe that interactive graphics also should enable users to explore the database underlying a map," write Stephen L. Egbert and Terry A. Slocum in, "EXPLOREMAP: An Exploration System for Choropleth Maps (*Annals of the Association of American Geographers*, 82(2), June, 1992, p. 275).

7. *Ibid.*

8. Gwendolyn Warren and William Bunge, *Field Notes: Discussion Paper No. 2: School Decentralization*, Detroit Geographical Expedition and Institute, Detroit, 1970, p. i.

9. *Ibid.*, p. 22.

11. W. Bunge, *Field Notes: Discussion Paper No. 1: The Detroit Geographical Expedition*, The Society for Human Exploration, Detroit, 1969, p. 38.

12. *Ibid.*

13. *Ibid.*

14. *Ibid.*, p. 39. As part of the output of the expeditions "an atlas to be entitled *The Atlas of Love and Hate* is also projected. A unique feature will be that each map will be the individual contribution of a particular geographer who will be asked to sign his map and brief essay, photographs and supplementary maps to create in a sense a journal of maps." Unfortunately this was never realized.

10. Bertin, *op. cit.*, p. 16. Peter Huber has recently taken up the question of scientific responsibility in the context of expert testimony (and what is a map if not expert testimony?). He closes, laudably but predictably enough, by rephasing Richard Feynman's counsel that, "A true scientist approached for advice on public matters must commit to come forward with his ultimate findings regardless of whose position they favor" ("Junk Science in the Courtroom," *Scientific American*, 266(6), June, 1992, p. 132). An easy way to do this is to publish *all the findings*. Feynman's faith was sorely tried in his final years as a result of his service on the commission investigating the *Challenger* disaster. So after observing, "The only way to have real success in science, the field I'm familiar with, is to describe the evidence very carefully without regard to the way you feel it should be. If you have a theory, you must try to explain what's good and what's bad about it equally. In science, you learn a kind of standard integrity and honesty," he goes on to say, "Now, Dr. Keel started out by telling me he had a degree in physics. I always assume that everybody in physics has integrity—perhaps I'm naive about that. . . ." (Richard Feynman, *"What Do You Care What Other People Think?"* Norton, New York, 1988, pp. 217–218). He was less naive than he makes out, but this faith in the integrity of science is as deep and abiding as the faith in the objectivity of the map.

15. Charles Hutchinson, John Carr and Laura Tangley, *BioDiversity at Risk: A Preview of Conservation International's Atlas for the 1990s*, Conservation International, Washington, D.C., 1992. The full color (plus varnish) 24" x 36" poster contains nine maps (though its own text miscounts these as seven)

reproduced from computer output, and essays about biodiversity, human disturbance, map production and conservation. It's pretty impressive.

16. *Ibid.*

17. What it says on the poster is: "The maps on this poster use the Peter's projection, which conveys the relative area of earth's land masses more accurately than the maps we are accustomed to seeing. Those maps, which use the Mercator Projection, exaggerate the size of Europe and North America relative to tropical Africa, Asia, and South America." The political implications of the choice in this context are especially obvious, or, uh, maybe they aren't. As Brian Harley wrote: "When I talk about the ideology of maps or ways of deconstructing the scientific claims of cartography I often think that I am either stating the obvious or tilting at windmills. I was rather reassured, therefore, by a tidbit from a recent Computer Network Conference. The discussion was about 'Choosing the Right World Atlas'. The contributor is Duane Marble, who had previously been a stranger to me. He states: 'A map projection is a mathematical transformation from three dimensional spherical coordinates (world) to two dimensional Cartesian coordinates (maps). *It escapes me how politics, etc. can enter into it.*' When I read this last sentence it was with a sense of disbelief. How many cartographers or GIS specialists actually suffer from tunnel vision? Do cartographers really believe anything they do is apolitical, that mapping is anything less than a form of empowerment, or that cartography lacks social and political consequences? These remarks are dedicated to Duane Marble and similar 'Marble hearted' mapmakers" (J. B. Harley, "What Happens When We Make a Map," unpublished lecture given at Pennsylvania State University, 1991, p. 4).

18. Hutchinson *et al., op. cit.*

19. Pradyumna P. Karan, "The Kingdom of Sikkim: Map Supplement No. 10," *Annals of the Association of American Geographers*, 59(1), March, 1969. The cartography is attributed to James Queen (the quite spectacular shaded relief to Eugene Zang). There was a period when the *Annals* published these egregiously pointless maps, skill lavished on skill, to no discernible end except that of displaying the skill required to make them. This map of Sikkim is thus a . . . *mapmaker's* . . . map. The acme (or nadir) of this practice would occur with the "Chico Demonstration Map" (Supplement to the *Annals of the Association of American Geographers*, 74(2), June, 1984) whose subtitle read, "Illustrating cartographic production principles and techniques in maps printed in from two to seven inks." As in Barthes' example of "*quia ego nominor leo*" from the Latin grammar, the ostensible subject and theme of this map have been *completely* consumed by the secondary semiological system (whereas in the map of Sikkim there is still the pretense that beyond—or below—the display of cartographic wizardry lies . . . a map of Sikkim).

20. Mark Monmonier, *How to Lie with Maps*, University of Chicago Press, Chicago, 1991, p. 123. Later in the same chapter Monmonier asks, "When a single variable might yield many different maps, which one is right? Or is this the key question? Should there be just one map? Should not the viewer be given several maps, or perhaps the opportunity to experiment with symbolization through a computer workstation?" The real problem is that Monmonier insists on construing the map user as a . . . *viewer* (that is, as a . . . *consumer*) instead of

as the mapmaking-decision-maker that Bertin, for example, imagines. Of course if the map user is a map*maker*, then there is no role for the cartographer. *¡Que lastima!*

21. Introductory note signed by the publishers of Ivan Illich, Irving Zola, John McKnight, Jonathan Kaplan and Harley Shaiken, *Disabling Professions*, Marion Boyars, London, 1977, p. 9.

22. Sarah Booth Conroy, "Your Friendly Neighborhood Map," *The Washington Post*, August 13, 1972, p. M1.

23. *Ibid*. Wiebenson's map was called, "The Adams-Morgan Map of Services". The brochure about his signage proposals bore the title, "A Public Information System for Adams-Morgan Streets."

24. Editor's note introducing Gregory Staple and Hugo Dixon, "Telegeography: Mapping the New World Order," *Whole Earth Review*, 75, Summer, 1992, pp. 124.

25. *Ibid*.

26. *Ibid*., p. 125.

27. To introduce the issue Bunge observes, "If the United States rolled a nuclear missile across the border from Minnesota into Ontario, Canadian customs, among others, would have a sovereign fit. If the same bomb were hoisted up 10 feet, Canada would be equally upset. If it were lifted 100 miles, it would not even be noticed. How high (in feet) is Canadian Sovereignty?" This leads to the imperative, "Readers must force themselves to think in terms of national boundary surfaces, not boundary lines." And this leads to the conclusion: "To state the problem in the exciting militaristic outlook of the 1990s, 'The Russians are not coming. They are already here.' They are not 'ninety miles off our shores' in Cuba or sitting in missile silos in the Soviet Union. They are up in the sky." Though these are all quoted from William Bunge, *Nuclear War Atlas*, Basil Blackwell, Oxford, 1988, pp. 71 and 83–85, the idea first appeared in William Bunge, *Patterns of Location: Michigan Inter-University Community of Mathematical Geographers Discussion Paper No. 3*, University of Michigan, Ann Arbor, February, 1964, p. 30.

28. The image appears in full color in Arthur Robinson, *Early Thematic Mapping in the History of Cartography*, University of Chicago Press, Chicago, 1982, p. 151; and in two colors in Edward Tufte, *The Visual Display of Quantitative Information*, Graphics Press, Cheshire, Connecticut, 1983, p. 25. Minard, a French engineer who pioneered a variety of statistical graphics, is a big favorite of Tufte's. This map puts France in the center of the world, pouring wine into the rest of Europe, North Africa, Argentina and Brazil.

29. This is another map reproduced by both Robinson, *op. cit.*, pp. 70–71; and Tufte, *op. cit.*, p. 23. A map of trade winds and monsoons, this is, in the end, another map of . . . *trade*.

30. A beautiful reproduction of a part of this appears as the frontispiece to P. D. A. Harvey's *Medieval Maps*, University of Toronto Press, Toronto, 1991, p. 2. Harvey's text here supports his general thesis about medieval European cartography, that it was an affair of sporadic and isolated efforts; but it also speaks of the connections maps make: "Few medieval people used or understood maps, and this itinerary, set out in graphic form by Matthew Paris, monk at St. Albans, in the mid-thirteenth century, was a highly original work. On this first of its five

pages, the left-hand strip shows the road from London, with its city walls and the medieval St. Paul's Cathedral, to Dover, with its castle. Crossing the English Channel, marked by waves and boats, the right-hand strip gives alternative routes from Calais and Boulogne to Reims and Beauvais" (p. 3).

31. A good close-up of part of this also appears in Harvey, *ibid.*, p. 6. Undermining the claims of the authors of the map of the telecom network to novelty in mapping network space, commentators have long noted that this fourth century map of the Roman Empire eschewed "geographic" space in the effort to produce a detailed and accurate guide to routes.

Index